naked statistics

also by Charles Wheelan

10½ Things No Commencement Speaker Has Ever Said

Naked Economics: Undressing the Dismal Science

naked statistics

Stripping the Dread from the Data

CHARLES WHEELAN

W. W. Norton & Company
New York | London

Printed in the United States of America
First Edition

For information about permission to reproduce selections from this book, write to
Permissions, W. W. Norton & Company, Inc.,
500 Fifth Avenue, New York, NY 10110

For information about special discounts for bulk purchases,
please contact W. W. Norton Special Sales at
specialsales@wwnorton.com or 800-233-4830

Manufacturing by Courier Westford
Production manager: Anna Oler

ISBN 978-0-393-07195-5 (hardcover)

W. W. Norton & Company, Inc.
500 Fifth Avenue, New York, N.Y. 10110
www.wwnorton.com

W. W. Norton & Company Ltd.
Castle House, 75/76 Wells Street, London W1T 3QT

3 4 5 6 7 8 9 0

For Katrina

Contents

Introduction

Why I hated calculus but love statistics

I have always had an uncomfortable relationship with math. I don't like numbers for the sake of numbers. I am not impressed by fancy formulas that have no real-world application. I particularly disliked high school calculus for the simple reason that no one ever bothered to tell me why I needed to learn it. What is the area beneath a parabola? Who cares?

In fact, one of the great moments of my life occurred during my senior year of high school, at the end of the first semester of Advanced Placement Calculus. I was working away on the final exam, admittedly less prepared for the exam than I ought to have been. (I had been accepted to my first-choice college a few weeks earlier, which had drained away what little motivation I had for the course.) As I stared at the final exam questions, they looked completely unfamiliar. I don't mean that I was having trouble answering the questions. I mean that I didn't even recognize what was being asked. I was no stranger to being unprepared for exams, but, to paraphrase Donald Rumsfeld, I usually knew what I didn't know. This exam looked even more Greek than usual. I flipped through the pages of the exam for a while and then more or less surrendered. I walked to the front of the classroom, where my calculus teacher, whom we'll call Carol Smith, was proctoring the exam. "Mrs. Smith," I said, "I don't recognize a lot of the stuff on the test."

Suffice it to say that Mrs. Smith did not like me a whole lot more than I liked her. Yes, I can now admit that I sometimes used my limited powers as student association president to schedule all-school assemblies just so that Mrs. Smith's calculus class would be canceled. Yes, my friends and I did have flowers delivered to Mrs. Smith during class from "a secret admirer" just so that we could chortle away in the back of the room as she looked around in embarrassment. And yes, I did stop doing any homework at all once I got in to college.

So when I walked up to Mrs. Smith in the middle of the exam and said that the material did not look familiar, she was, well, unsympathetic. "Charles," she said loudly, ostensibly to me but facing the rows of desks to make certain that the whole class could hear, "if you had studied, the material would look a lot more familiar." This was a compelling point.

So I slunk back to my desk. After a few minutes, Brian Arbetter, a far better calculus student than I, walked to the front of the room and whispered a few things to Mrs. Smith. She whispered back and then a truly extraordinary thing happened. "Class, I need your attention," Mrs. Smith announced. "It appears that I have given you the second semester exam by mistake." We were far enough into the test period that the whole exam had to be aborted and rescheduled.

I cannot fully describe my euphoria. I would go on in life to marry a wonderful woman. We have three healthy children. I've published books and visited places like the Taj Mahal and Angkor Wat. Still, the day that my calculus teacher got her comeuppance is a top five life moment. (The fact that I nearly failed the makeup final exam did not significantly diminish this wonderful life experience.)

The calculus exam incident tells you much of what you need to know about my relationship with mathematics—but not everything. Curiously, I loved physics in high school, even though physics relies very heavily on the very same calculus that I refused to do in Mrs. Smith's class. Why? *Because physics has a clear purpose.* I distinctly remember my high school physics teacher showing us during the World Series how we could use the basic formula for acceleration to estimate how far a home run had been hit. That's cool—and the same formula has many more socially significant applications.

Once I arrived in college, I thoroughly enjoyed probability, again because it offered insight into interesting real-life situations. In hindsight, I now recognize that it wasn't the math that bothered me in calculus class; it was that no one ever saw fit to explain the point of it. If you're not fascinated by the elegance of formulas alone—which I am most emphatically not—then it is just a lot of tedious and mechanistic formulas, at least the way it was taught to me.

That brings me to statistics (which, for the purposes of this book, includes probability). I love statistics. Statistics can be used to explain everything from DNA testing to the idiocy of playing the lottery. Statistics can help us identify the factors associated with diseases like cancer and heart disease; it can help us spot cheating on standardized tests. Statistics can even help you win on game shows. There was a famous program during my childhood called *Let's Make a Deal*, with its equally famous host, Monty Hall. At the end of each day's show, a successful player would stand with Monty facing three big doors: Door no. 1, Door no. 2, and Door no. 3. Monty Hall explained to the player that there was a highly desirable prize behind one of the doors—something like a new car—and a goat behind the other two. The idea was straightforward: the player chose one of the doors and would get the contents behind that door.

As each player stood facing the doors with Monty Hall, he or she had a 1 in 3 chance of choosing the door that would be opened to reveal the valuable prize. But *Let's Make a Deal* had a twist, which has delighted statisticians ever since (and perplexed everyone else). After the player chose a door, Monty Hall would open one of the two remaining doors, always revealing a goat. For the sake of example, assume that the player has chosen Door no. 1. Monty would then open Door no. 3; the live goat would be standing there on stage. Two doors would still be closed, nos. 1 and 2. If the valuable prize was behind no. 1, the contestant would win; if it was behind no. 2, he would lose. But then things got more interesting: Monty would turn to the player and ask whether he would like to change his mind and switch doors (from no. 1 to no. 2 in this case). Remember, both doors were still closed, and the only new information the contestant had received was that a goat showed up behind one of the doors that he didn't pick.

Should he switch?

The answer is yes. Why? That's in Chapter 5½.

The paradox of statistics is that they are everywhere—from batting averages to presidential polls—but the discipline itself has a reputation for being uninteresting and inaccessible. Many statistics books and classes are overly laden with math and jargon. Believe me, the technical details are crucial (and interesting)—but it's just Greek if you don't understand the intuition. And you may not even care about the intuition if you're not convinced that there is any reason to learn it. Every chapter in this book promises to answer the basic question that I asked (to no effect) of my high school calculus teacher: *What is the point of this?*

This book is about the intuition. It is short on math, equations, and graphs; when they are used, I promise that they will have a clear and enlightening purpose. Meanwhile, the book is long on examples to convince you that there are great reasons to learn this stuff. *Statistics can be really interesting, and most of it isn't that difficult.*

The idea for this book was born not terribly long after my unfortunate experience in Mrs. Smith's AP Calculus class. I went to graduate school to study economics and public policy. Before the program even started, I was assigned (not surprisingly) to "math camp" along with the bulk of my classmates to prepare us for the quantitative rigors that were to follow. For three weeks, we learned math all day in a windowless, basement classroom (really).

On one of those days, I had something very close to a career epiphany. Our instructor was trying to teach us the circumstances under which the sum of an infinite series converges to a finite number. Stay with me here for a minute because this concept will become clear. (Right now you're probably feeling the way I did in that windowless classroom.) An infinite series is a pattern of numbers that goes on forever, such as $1 + \frac{1}{2} + \frac{1}{4} + \frac{1}{8} \ldots$ The three dots means that the pattern continues to infinity.

This is the part we were having trouble wrapping our heads around. Our instructor was trying to convince us, using some proof I've long since forgotten, that a series of numbers can go on forever and yet still add up (roughly) to a finite number. One of my classmates, Will Warshauer, would have none of it, despite the impressive mathematical proof. (To be honest, I was a bit skeptical myself.) How can something that is infinite add up to something that is finite?

Then I got an inspiration, or more accurately, the intuition of what the instructor was trying to explain. I turned to Will and talked him through what I had just worked out in my head. Imagine that you have positioned yourself exactly 2 feet from a wall.

Now move half the distance to that wall (1 foot), so that you are left standing 1 foot away.

From 1 foot away, move half the distance to the wall once again (6 inches, or ½ a foot). And from 6 inches away, do it again (move 3 inches, or ¼ of a foot). Then do it again (move 1½ inches, or ⅛ of a foot). And so on.

You will gradually get pretty darn close to the wall. (For example, when you are 1/1024th of an inch from the wall, you will move half the distance, or another 1/2048th of an inch.) But you will never hit the wall, because by definition each move takes you only half the remaining distance. In other words, you will get infinitely close to the wall but never hit it. If we measure your moves in feet, the series can be described as 1 + ½ + ¼ + ⅛ . . .

Therein lies the insight: Even though you will continue moving forever—with each move taking you half the remaining distance to the wall—the total distance you travel can never be more than 2 feet, which is your starting distance from the wall. For mathematical purposes, the total distance you travel can be approximated as 2 feet, which turns out to be very handy for computation purposes. A mathematician would say that the sum of this infinite series 1 ft + ½ ft + ¼ ft + ⅛ ft . . . converges to 2 feet, which is what our instructor was trying to teach us that day.

The point is that I convinced Will. I convinced myself. I can't remember the math proving that the sum of an infinite series can converge to a finite number, but I can always look that up online. And when I do, it will probably make sense. In my experience, the intuition makes the math and other technical details more understandable—but not necessarily the other way around.

The point of this book is to make the most important statistical concepts more intuitive and more accessible, not just for those of us forced to study them in windowless classrooms but for anyone interested in the extraordinary power of numbers and data.

Now, having just made the case that the core tools of statistics are less intuitive and accessible than they ought to be, I'm going to make a seemingly contradictory point: Statistics can be *overly accessible* in the sense that anyone with data and a computer can do sophisticated statistical procedures with a few keystrokes. The problem is that if the data are poor, or if the statistical techniques are used improperly, the conclusions can be wildly misleading and even potentially dangerous. Consider the following hypothetical Internet news flash: *People Who Take Short Breaks at Work Are Far More Likely to Die of Cancer.* Imagine that headline popping up while you are surfing the Web. According to a seemingly impressive study of 36,000 office workers (a huge data set!), those workers who reported leaving their offices to take regular ten-minute breaks during the workday were 41 percent more likely to develop cancer over the next five years than workers who don't leave their offices during the workday. Clearly we need to act on this kind of finding—perhaps some kind of national awareness campaign to prevent short breaks on the job.

Or maybe we just need to think more clearly about what many workers are doing during that ten-minute break. My professional experience suggests that many of those workers who report leaving their offices for short breaks are huddled outside the entrance of the building smoking cigarettes (creating a haze of smoke through which the rest of us have to walk in order to get in or out). I would further infer that it's probably the cigarettes, and not the short breaks from work, that are causing the cancer. I've made up this example just so that it would be particularly absurd, but I can assure you that many real-life statistical abominations are nearly this absurd once they are deconstructed.

Statistics is like a high-caliber weapon: helpful when used correctly and potentially disastrous in the wrong hands. This book *will not* make you a statistical expert; it *will* teach you enough care and respect for the field that you don't do the statistical equivalent of blowing someone's head off.

This is not a textbook, which is liberating in terms of the topics that have to be covered and the ways in which they can be explained. *The book has been designed to introduce the statistical concepts with the most relevance to everyday life*. How do scientists conclude that something causes cancer? How does polling work (and what can go wrong)? Who "lies with statistics," and how do they do it? How does your credit card company use data on what you are buying to predict if you are likely to miss a payment? (Seriously, they can do that.)

If you want to understand the numbers behind the news and to appreciate the extraordinary (and growing) power of data, this is the stuff you need to know. In the end, I hope to persuade you of the observation first made by Swedish mathematician and writer Andrejs Dunkels: It's easy to lie with statistics, but it's hard to tell the truth without them.

But I have even bolder aspirations than that. I think you might actually enjoy statistics. The underlying ideas are fabulously interesting and relevant. The key is to separate the important ideas from the arcane technical details that can get in the way. That is Naked Statistics.

acknowledgments

This book was conceived as an homage to an earlier W. W. Norton classic, *How to Lie with Statistics* by Darrell Huff, which was written in the 1950s and has sold over a million copies. That book, like this one, was written to demystify statistics and persuade everyday readers that what they don't understand about the numbers behind the headlines *can* hurt them. I hope that I've done justice to Mr. Huff's classic. In any event, I would be delighted to have sold a million copies fifty years from now!

I am continually grateful to W. W. Norton, and to Drake McFeely in particular, for enabling me to write books that address significant topics in a way that is understandable to lay readers. Drake has been a great friend and supporter for more than a decade now.

Jeff Shreve is the guy at W. W. Norton who brought this book to fruition. Upon meeting Jeff, one might think that he is far too nice to enforce the multiple deadlines involved with the production of a book like this. Not true. Yes, he really is that nice, but somehow his gentle prodding seems to get the job done. (For example, these acknowledgments are due tomorrow morning.) I appreciate having a kind taskmaster to move things along.

My biggest debt of gratitude goes to the many men and women who do the important research and analysis described in this book. I am not a statistician, nor am I a researcher. I am merely a translator of other people's interesting and significant work. I hope that I have conveyed throughout this book how important good research and sound analysis are to making us healthier, wealthier, safer, and better informed.

In particular, I would like to acknowledge the wide-ranging work of Princeton economist Alan Krueger, who has made clever and significant research contributions on topics ranging from the roots of terrorism to the economic returns from higher education. (His findings on both of these topics are pleasantly counterintuitive.) More importantly (to me), Alan was one of my graduate school statistics professors; I have always been impressed by his ability to successfully balance research, teaching, and public service.

Jim Sallee, Jeff Grogger, Patty Anderson, and Arthur Minetz all read earlier drafts of the manuscript and made numerous helpful suggestions. Thank you for saving me from myself! Frank Newport of Gallup and Mike Kagay from the *New York Times* were kind enough to spend time walking me through the methodological nuances of polling. Despite all of their efforts, the mistakes that remain are my own.

Katie Wade was an indefatigable research assistant. (I have always wanted to use the word "indefatigable" and, finally, this is the perfect context.) Katie is the source of many of the anecdotes and examples that illuminate concepts throughout the book. No Katie, no fun examples.

I have wanted to write books since I was in elementary school. The person who enables me to do that, and to make a living at it, is my agent, Tina Bennett. Tina embodies the best of the publishing business. She delights in bringing meaningful work to fruition while tirelessly promoting the interests of her clients.

And last, my family deserves credit for tolerating me as I produced this book. (The chapter deadlines were posted on the refrigerator.) There is evidence that I become 31 percent crankier and 23 percent more exhausted when approaching (or missing) major book deadlines. My wife, Leah, is the first, best, and most important editor of everything that I write. Thank you for that, and for being such a smart, supportive, and fun partner in all other endeavors.

The book is dedicated to my oldest daughter, Katrina. It is hard to believe that the child who was in a crib when I wrote *Naked Economics* can now read chapters and provide meaningful feedback. Katrina, you are a parent's dream, as are Sophie and C.J., who will soon be reading chapters and manuscripts, too.

naked statistics

CHAPTER 1

What's the Point?

I've noticed a curious phenomenon. Students will complain that statistics is confusing and irrelevant. Then the same students will leave the classroom and happily talk over lunch about batting averages (during the summer) or the windchill factor (during the winter) or grade point averages (always). They will recognize that the National Football League's "passer rating"—a statistic that condenses a quarterback's performance into a single number—is a somewhat flawed and arbitrary measure of a quarterback's game day performance. The same data (completion rate, average yards per pass attempt, percentage of touchdown passes per pass attempt, and interception rate) could be combined in a different way, such as giving greater or lesser weight to any of those inputs, to generate a different but equally credible measure of performance. Yet anyone who has watched football recognizes that it's handy to have a single number that can be used to encapsulate a quarterback's performance.

Is the quarterback rating perfect? No. Statistics rarely offers a single "right" way of doing anything. Does it provide meaningful information in an easily accessible way? Absolutely. It's a nice tool for making a quick comparison between the performances of two quarterbacks on a given day. I am a Chicago Bears fan. During the 2011 playoffs, the Bears played the Packers; the Packers won. There are a lot of ways I could describe that game, including pages and pages of analysis and raw data. But here is a more succinct analysis. Chicago Bears quarterback Jay Cutler had a passer rating of 31.8. In contrast, Green Bay quarterback Aaron Rodgers had a passer rating of 55.4. Similarly, we can compare Jay Cutler's performance to that in a game earlier in the season against Green Bay, when he had a passer rating of 85.6. That tells you a lot of what you need to know in order to understand why the Bears beat the Packers earlier in the season but lost to them in the playoffs.

That is a very helpful synopsis of what happened on the field. Does it simplify things? Yes, that is both the strength and the weakness of any descriptive statistic. One number tells you that Jay Cutler was outgunned by Aaron Rodgers in the Bears' playoff loss. On the other hand, that number won't tell you whether a quarterback had a bad break, such as throwing a perfect pass that was bobbled by the receiver and then intercepted, or whether he "stepped up" on certain key plays (since every completion is weighted the same, whether it is a crucial third down or a meaningless play at the end of the game), or whether the defense was terrible. And so on.

The curious thing is that the same people who are perfectly comfortable discussing statistics in the context of sports or the weather or grades will seize up with anxiety when a researcher starts to explain something like the Gini index, which is a standard tool in economics for measuring income inequality. I'll explain what the Gini index is in a moment, but for now *the most important thing to recognize is that the Gini index is just like the passer rating*. It's a handy tool for collapsing complex information into a single number. As such, it has the strengths of most descriptive statistics, namely that it provides an easy way to compare the income distribution in two countries, or in a single country at different points in time.

The Gini index measures how evenly wealth (or income) is shared within a country on a scale from zero to one. The statistic can be calculated for wealth or for annual income, and it can be calculated at the individual level or at the household level. (All of these statistics will be highly correlated but not identical.) The Gini index, like the passer rating, has no intrinsic meaning; it's a tool for comparison. A country in which every household had identical wealth would have a Gini index of zero. By contrast, a country in which a single household held the country's entire wealth would have a Gini index of one. As you can probably surmise, the closer a country is to one, the more unequal its distribution of wealth. The United States has a Gini index of .45, according to the Central Intelligence Agency (a great collector of statistics, by the way).[1] So what?

Once that number is put into context, it can tell us a lot. For example, Sweden has a Gini index of .23. Canada's is .32. China's is .42. Brazil's is .54. South Africa's is .65.* As we look across those numbers, we get a sense of where the United States falls relative to the rest of the world when it comes to income inequality. We can also compare different points in time. The Gini index for the United States was .41 in 1997 and grew to .45 over the next decade. (The most recent CIA data are for 2007.) This tells us in an objective way that while the United States grew richer over that period of time, the distribution of wealth grew more unequal. Again, we can compare the changes in the Gini index across countries over roughly the same time period. Inequality in Canada was basically unchanged over the same stretch. Sweden has had significant economic growth over the past two decades, but the Gini index in Sweden actually fell from .25 in 1992 to .23 in 2005, meaning that Sweden grew richer *and* more equal over that period.

Is the Gini index the perfect measure of inequality? Absolutely not—just as the passer rating is not a perfect measure of quarterback performance. But it certainly gives us some valuable information on a socially significant phenomenon in a convenient format.

We have also slowly backed our way into answering the question posed in the chapter title: What is the point? The point is that statistics helps us process data, which is really just a fancy name for information. Sometimes the data are trivial in the grand scheme of things, as with sports statistics. Sometimes they offer insight into the nature of human existence, as with the Gini index.

---

* The Gini index is sometimes multiplied by 100 to make it a whole number. In that case, the United States would have a Gini Index of 45.

But, as any good infomercial would point out, *That's not all!* Hal Varian, chief economist at Google, told the *New York Times* that being a statistician will be "the sexy job" over the next decade.[2] I'll be the first to concede that economists sometimes have a warped definition of "sexy." Still, consider the following disparate questions:

How can we catch schools that are cheating on their standardized tests?

How does Netflix know what kind of movies you like?

How can we figure out what substances or behaviors cause cancer, given that we cannot conduct cancer-causing experiments on humans?

Does praying for surgical patients improve their outcomes?

Is there really an economic benefit to getting a degree from a highly selective college or university?

What is causing the rising incidence of autism?

Statistics can help answer these questions (or, we hope, can soon). The world is producing more and more data, ever faster and faster. Yet, as the *New York Times* has noted, "Data is merely the raw material of knowledge."[3]* Statistics is the most powerful tool we have for using information to some meaningful end, whether that is identifying underrated baseball players or paying teachers more fairly. Here is a quick tour of how statistics can bring meaning to raw data.

Description and Comparison

A bowling score is a descriptive statistic. So is a batting average. Most American sports fans over the age of five are already conversant in the

* The word "data" has historically been considered plural (e.g., "The data are very encouraging.") The singular is "datum," which would refer to a single data point, such as one person's response to a single question on a poll. Using the word "data" as a plural noun is a quick way to signal to anyone who does serious research that you are conversant with statistics. That said, many authorities on grammar and many publications, such as the *New York Times*, now accept that "data" can be singular or plural, as the passage that I've quoted from the *Times* demonstrates.

field of descriptive statistics. We use numbers, in sports and everywhere else in life, to summarize information. How good a baseball player was Mickey Mantle? He was a career .298 hitter. To a baseball fan, that is a meaningful statement, which is remarkable when you think about it, because it encapsulates an eighteen-season career.[4] (There is, I suppose, something mildly depressing about having one's lifework collapsed into a single number.) Of course, baseball fans have also come to recognize that descriptive statistics other than batting average may better encapsulate a player's value on the field.

We evaluate the academic performance of high school and college students by means of a grade point average, or GPA. A letter grade is assigned a point value; typically an A is worth 4 points, a B is worth 3, a C is worth 2, and so on. By graduation, when high school students are applying to college and college students are looking for jobs, the grade point average is a handy tool for assessing their academic potential. Someone who has a 3.7 GPA is clearly a stronger student than someone at the same school with a 2.5 GPA. That makes it a nice descriptive statistic. It's easy to calculate, it's easy to understand, and it's easy to compare across students.

But it's not perfect. The GPA does not reflect the difficulty of the courses that different students may have taken. How can we compare a student with a 3.4 GPA in classes that appear to be relatively nonchallenging and a student with a 2.9 GPA who has taken calculus, physics, and other tough subjects? I went to a high school that attempted to solve this problem by giving extra weight to difficult classes, so that an A in an "honors" class was worth five points instead of the usual four. This caused its own problems. My mother was quick to recognize the distortion caused by this GPA "fix." For a student taking a lot of honors classes (me), any A in a nonhonors course, such as gym or health education, would actually pull my GPA down, even though it is impossible to do better than an A in those classes. As a result, my parents forbade me to take driver's education in high school, lest even a perfect performance diminish my chances of getting into a competitive college and going on to write popular books. Instead, they paid to send me to a private driving school, at nights over the summer.

Was that insane? Yes. But one theme of this book will be that an overreliance on any descriptive statistic can lead to misleading conclusions, or cause undesirable behavior. My original draft of that sentence used the phrase "oversimplified descriptive statistic," but I struck the word "oversimplified" because it's redundant. Descriptive statistics exist to simplify, which always implies some loss of nuance or detail. Anyone working with numbers needs to recognize as much.

Inference

How many homeless people live on the streets of Chicago? How often do married people have sex? These may seem like wildly different kinds of questions; in fact, they both can be answered (not perfectly) by the use of basic statistical tools. One key function of statistics is to use the data we have to make informed conjectures about larger questions for which we do not have full information. In short, we can use data from the "known world" to make informed inferences about the "unknown world."

Let's begin with the homeless question. It is expensive and logistically difficult to count the homeless population in a large metropolitan area. Yet it is important to have a numerical estimate of this population for purposes of providing social services, earning eligibility for state and federal revenues, and gaining congressional representation. One important statistical practice is sampling, which is the process of gathering data for a small area, say, a handful of census tracts, and then using those data to make an informed judgment, or inference, about the homeless population for the city as a whole. Sampling requires far less resources than trying to count an entire population; done properly, it can be every bit as accurate.

A political poll is one form of sampling. A research organization will attempt to contact a sample of households that are broadly representative of the larger population and ask them their views about a particular issue or candidate. This is obviously much cheaper and faster than trying to contact every household in an entire state or country. The polling and research firm Gallup reckons that a methodologically sound poll of 1,000 households will produce roughly the same results as a poll that attempted to contact every household in America.

That's how we figured out how often Americans are having sex, with whom, and what kind. In the mid-1990s, the National Opinion Research Center at the University of Chicago carried out a remarkably ambitious study of American sexual behavior. The results were based on detailed surveys conducted in person with a large, representative sample of American adults. If you read on, Chapter 10 will tell you what they learned. *How many other statistics books can promise you that?*

Assessing Risk and Other Probability-Related Events

Casinos make money in the long run—always. That does not mean that they are making money at any given moment. When the bells and whistles go off, some high roller has just won thousands of dollars. The whole gambling industry is built on games of chance, meaning that the outcome of any particular roll of the dice or turn of the card is uncertain. At the same time, the underlying probabilities for the relevant events—drawing 21 at blackjack or spinning red in roulette—are known. When the underlying probabilities favor the casinos (as they always do), we can be increasingly certain that the "house" is going to come out ahead as the number of bets wagered gets larger and larger, even as those bells and whistles keep going off.

This turns out to be a powerful phenomenon in areas of life far beyond casinos. Many businesses must assess the risks associated with assorted adverse outcomes. They cannot make those risks go away entirely, just as a casino cannot guarantee that you won't win every hand of blackjack that you play. However, any business facing uncertainty can manage these risks by engineering processes so that the probability of an adverse outcome, anything from an environmental catastrophe to a defective product, becomes acceptably low. Wall Street firms will often evaluate the risks posed to their portfolios under different scenarios, with each of those scenarios weighted based on its probability. The financial crisis of 2008 was precipitated in part by a series of market events that had been deemed extremely unlikely, as if every player in a casino drew blackjack all night. I will argue later in the book that these Wall Street models were flawed and that the data they used to assess the underlying risks were too limited, but the point here is that any model to deal with risk must have probability as its foundation.

When individuals and firms cannot make unacceptable risks go away, they seek protection in other ways. The entire insurance industry is built upon charging customers to protect them against some adverse outcome, such as a car crash or a house fire. The insurance industry does not make money by eliminating these events; cars crash and houses burn every day. Sometimes cars even crash into houses, causing them to burn. Instead, the insurance industry makes money by charging premiums that are more than sufficient to pay for the expected payouts from car crashes and house fires. (The insurance company may also try to lower its expected payouts by encouraging safe driving, fences around swimming pools, installation of smoke detectors in every bedroom, and so on.)

Probability can even be used to catch cheats in some situations. The firm Caveon Test Security specializes in what it describes as "data forensics" to find patterns that suggest cheating.[5] For example, the company (which was founded by a former test developer for the SAT) will flag exams at a school or test site on which the number of identical *wrong answers* is highly unlikely, usually a pattern that would happen by chance less than one time in a million. The mathematical logic stems from the fact that we cannot learn much when a large group of students all answer a question correctly. That's what they are supposed to do; they could be cheating, or they could be smart. But when those same test takers get an answer wrong, they should not all consistently have *the same wrong answer*. If they do, it suggests that they are copying from one another (or sharing answers via text). The company also looks for exams in which a test taker does significantly better on hard questions than on easy questions (suggesting that he or she had answers in advance) and for exams on which the number of "wrong to right" erasures is significantly higher than the number of "right to wrong" erasures (suggesting that a teacher or administrator changed the answer sheets after the test).

Of course, you can see the limitations of using probability. A large group of test takers might have the same wrong answers by coincidence; in fact, the more schools we evaluate, the more likely it is that we will observe such patterns just as a matter of chance. A statistical anomaly does not prove wrongdoing. Delma Kinney, a fifty-year-old Atlanta man, won $1 million in an instant lottery game in 2008 and then another $1 million in an instant game in 2011.[6] The probability of that happening to the same person is somewhere in the range of 1 in 25 trillion. We cannot arrest Mr. Kinney for fraud on the basis of that calculation alone (though we might inquire whether he has any relatives who work for the state lottery). Probability is one weapon in an arsenal that requires good judgment.

Identifying Important Relationships (Statistical Detective Work)

Does smoking cigarettes cause cancer? We have an answer for that question—but the process of answering it was not nearly as straightforward as one might think. The scientific method dictates that if we are testing a scientific hypothesis, we should conduct a controlled experiment in which the variable of interest (e.g., smoking) is the only thing that differs between the experimental group and the control group. If we observe a marked difference in some outcome between the two groups (e.g., lung cancer), we can safely infer that the variable of interest is what caused that outcome. We cannot do that kind of experiment on humans. If our working hypothesis is that smoking causes cancer, it would be unethical to assign recent college graduates to two groups, smokers and nonsmokers, and then see who has cancer at the twentieth reunion. (We can conduct controlled experiments on humans when our hypothesis is that a new drug or treatment may improve their health; we cannot knowingly expose human subjects when we expect an adverse outcome.)*

---

* This is a gross simplification of the fascinating and complex field of medical ethics.

Now, you might point out that we do not need to conduct an ethically dubious experiment to observe the effects of smoking. Couldn't we just skip the whole fancy methodology and compare cancer rates at the twentieth reunion between those who have smoked since graduation and those who have not?

No. Smokers and nonsmokers are likely to be different in ways other than their smoking behavior. For example, smokers may be more likely to have other habits, such as drinking heavily or eating badly, that cause adverse health outcomes. If the smokers are particularly unhealthy at the twentieth reunion, we would not know whether to attribute this outcome to smoking or to other unhealthy things that many smokers happen to do. We would also have a serious problem with the data on which we are basing our analysis. Smokers who have become seriously ill with cancer are less likely to attend the twentieth reunion. (The dead smokers definitely won't show up.) As a result, any analysis of the health of the attendees at the twentieth reunion (related to smoking or anything else) will be seriously flawed by the fact that the healthiest members of the class are the most likely to show up. The further the class gets from graduation, say, a fortieth or a fiftieth reunion, the more serious this bias will be.

We cannot treat humans like laboratory rats. As a result, statistics is a lot like good detective work. The data yield clues and patterns that can ultimately lead to meaningful conclusions. You have probably watched one of those impressive police procedural shows like *CSI: New York* in which very attractive detectives and forensic experts pore over minute clues—DNA from a cigarette butt, teeth marks on an apple, a single fiber from a car floor mat—and then use the evidence to catch a violent criminal. The appeal of the show is that these experts do not have the conventional evidence used to find the bad guy, such as an eyewitness or a surveillance videotape. So they turn to scientific inference instead. Statistics does basically the same thing. The data present unorganized clues—the crime scene. Statistical analysis is the detective work that crafts the raw data into some meaningful conclusion.

After Chapter 11, you will appreciate the television show I hope to pitch: *CSI: Regression Analysis*, which would be only a small departure from those other action-packed police procedurals. Regression analysis is the tool that enables researchers to isolate a relationship between two variables, such as smoking and cancer, while holding constant (or "controlling for") the effects of other important variables, such as diet, exercise, weight, and so on. When you read in the newspaper that eating a bran muffin every day will reduce your chances of getting colon cancer, you need not fear that some unfortunate group of human experimental subjects has been force-fed bran muffins in the basement of a federal laboratory somewhere while the control group in the next building gets bacon and eggs. Instead, researchers will gather detailed information on thousands of people, including how frequently they eat bran muffins, and then use regression analysis to do two crucial things: (1) quantify the association observed between eating bran muffins and contracting colon cancer (e.g., a hypothetical finding that people who eat bran muffins have a 9 percent lower incidence of colon cancer, controlling for other factors that may affect the incidence of the disease); and (2) quantify the likelihood that the association between bran muffins and a lower rate of colon cancer observed in this study is merely a coincidence—a quirk in the data for this sample of people—rather than a meaningful insight about the relationship between diet and health.

Of course, *CSI: Regression Analysis* will star actors and actresses who are much better looking than the academics who typically pore over such data. These hotties (all of whom would have PhDs, despite being only twenty-three years old) would study large data sets and use the latest statistical tools to answer important social questions: What are the most effective tools for fighting violent crime? What individuals are most likely to become terrorists? Later in the book we will discuss the concept of a "statistically significant" finding, which means that the analysis has uncovered an association between two variables that is not likely to be the product of chance alone. For academic researchers, this kind of statistical finding is the "smoking gun." On *CSI: Regression Analysis*, I envision a researcher working late at night in the computer lab because of her daytime commitment as a member of the U.S. Olympic beach volleyball team. When she gets the printout from her statistical analysis, she sees exactly what she has been looking for: a large and statistically significant relationship in her data set between some variable that she had hypothesized might be important and the onset of autism. She must share this breakthrough immediately!

The researcher takes the printout and runs down the hall, slowed somewhat by the fact that she is wearing high heels and a relatively small, tight black skirt. She finds her male partner, who is inexplicably fit and tan for a guy who works fourteen hours a day in a basement computer lab, and shows him the results. He runs his fingers through his neatly trimmed goatee, grabs his Glock 9-mm pistol from the desk drawer, and slides it into the shoulder holster beneath his $5,000 Hugo Boss suit (also inexplicable given his starting academic salary of $38,000 a year). Together the regression analysis experts walk briskly to see their boss, a grizzled veteran who has overcome failed relationships and a drinking problem . . .

Okay, you don't have to buy into the television drama to appreciate the importance of this kind of statistical research. Just about every social challenge that we care about has been informed by the systematic analysis of large data sets. (In many cases, gathering the relevant data, which is expensive and time-consuming, plays a crucial role in this process as will be explained in Chapter 7.) I may have embellished my characters in *CSI: Regression Analysis* but not the kind of significant questions they could examine. There is an academic literature on terrorists and suicide bombers—a subject that would be difficult to study by means of human subjects (or lab rats for that matter). One such book, *What Makes a Terrorist*, was written by one of my graduate school statistics professors. The book draws its conclusions from data gathered on terrorist attacks around the world. A sample finding: Terrorists are not desperately poor, or poorly educated. The author, Princeton economist Alan Krueger, concludes, "Terrorists tend to be drawn from well-educated, middle-class or high-income families."[7]

Why? Well, that exposes one of the limitations of regression analysis. We can isolate a strong association between two variables by using statistical analysis, but we cannot necessarily explain why that relationship exists, and in some cases, we cannot know for certain that the relationship is causal, meaning that a change in one variable is really causing a change in the other. In the case of terrorism, Professor Krueger hypothesizes that since terrorists are motivated by political goals, those who are most educated and affluent have the strongest incentive to change society. These individuals may also be particularly rankled by suppression of freedom, another factor associated with terrorism. In Krueger's study, countries with high levels of political repression have more terrorist activity (holding other factors constant).

This discussion leads me back to the question posed by the chapter title: What is the point? The point is not to do math, or to dazzle friends and colleagues with advanced statistical techniques. The point is to learn things that inform our lives.

Lies, Damned Lies, and Statistics

Even in the best of circumstances, statistical analysis rarely unveils "the truth." We are usually building a circumstantial case based on imperfect data. As a result, there are numerous reasons that intellectually honest individuals may disagree about statistical results or their implications. At the most basic level, we may disagree on the question that is being answered. Sports enthusiasts will be arguing for all eternity over "the best baseball player ever" because there is no objective definition of "best." Fancy descriptive statistics can inform this question, but they will never answer it definitively. As the next chapter will point out, more socially significant questions fall prey to the same basic challenge. What is happening to the economic health of the American middle class? That answer depends on how one defines both "middle class" and "economic health."

There are limits on the data we can gather and the kinds of experiments we can perform. Alan Krueger's study of terrorists did not follow thousands of youth over multiple decades to observe which of them evolved into terrorists. It's just not possible. Nor can we create two identical nations—except that one is highly repressive and the other is not—and then compare the number of suicide bombers that emerge in each. Even when we can conduct large, controlled experiments on human beings, they are neither easy nor cheap. Researchers did a large-scale study on whether or not prayer reduces postsurgical complications, which was one of the questions raised earlier in this chapter. *That study cost $2.4 million.* (For the results, you'll have to wait until Chapter 13.)

Secretary of Defense Donald Rumsfeld famously said, "You go to war with the army you have—not the army you might want or wish to have at a later time." Whatever you may think of Rumsfeld (and the Iraq war that he was explaining), that aphorism applies to research, too. We conduct statistical analysis using the best data and methodologies and resources available. The approach is not like addition or long division, in which the correct technique yields the "right" answer and a computer is always more precise and less fallible than a human. Statistical analysis is more like good detective work (hence the commercial potential of *CSI: Regression Analysis*). Smart and honest people will often disagree about what the data are trying to tell us.

But who says that everyone using statistics is smart or honest? As mentioned, this book began as an homage to *How to Lie with Statistics*, which was first published in 1954 and has sold over a million copies. The reality is that you *can* lie with statistics. Or you can make inadvertent errors. In either case, the mathematical precision attached to statistical analysis can dress up some serious nonsense. This book will walk through many of the most common statistical errors and misrepresentations (so that you can recognize them, not put them to use).

So, to return to the title chapter, what is the point of learning statistics?

To summarize huge quantities of data.

To make better decisions.

To answer important social questions.

To recognize patterns that can refine how we do everything from selling diapers to catching criminals.

To catch cheaters and prosecute criminals.

To evaluate the effectiveness of policies, programs, drugs, medical procedures, and other innovations.

And to spot the scoundrels who use these very same powerful tools for nefarious ends.

If you can do all of that while looking great in a Hugo Boss suit or a short black skirt, then you might also be the next star of *CSI: Regression Analysis*.

CHAPTER 2

Descriptive Statistics

Who was the best baseball player of all time?

Let us ponder for a moment two seemingly unrelated questions: (1) What is happening to the economic health of America's middle class? and (2) Who was the greatest baseball player of all time?

The first question is profoundly important. It tends to be at the core of presidential campaigns and other social movements. The middle class is the heart of America, so the economic well-being of that group is a crucial indicator of the nation's overall economic health. The second question is trivial (in the literal sense of the word), but baseball enthusiasts can argue about it endlessly. What the two questions have in common is that they can be used to illustrate the strengths and limitations of descriptive statistics, which are the numbers and calculations we use to summarize raw data.

If I want to demonstrate that Derek Jeter is a great baseball player, I can sit you down and describe every at bat in every Major League game that he's played. That would be raw data, and it would take a while to digest, given that Jeter has played seventeen seasons with the New York Yankees and taken 9,868 at bats.

Or I can just tell you that at the end of the 2011 season Derek Jeter had a career batting average of .313. That is a descriptive statistic, or a "summary statistic."

The batting average is a gross simplification of Jeter's seventeen seasons. It is easy to understand, elegant in its simplicity—and limited in what it can tell us. Baseball experts have a bevy of descriptive statistics that they consider to be more valuable than the batting average. I called Steve Moyer, president of Baseball Info Solutions (a firm that provides a lot of the raw data for the *Moneyball* types), to ask him, (1) What are the most important statistics for evaluating baseball talent? and (2) Who was the greatest player of all time? I'll share his answer once we have more context.

Meanwhile, let's return to the less trivial subject, the economic health of the middle class. Ideally we would like to find the economic equivalent of a batting average, or something even better. We would like a simple but accurate measure of how the economic well-being of the typical American worker has been changing in recent years. Are the people we define as middle class getting richer, poorer, or just running in place? A reasonable answer—though by no means the "right" answer—would be to calculate the change in per capita income in the United States over the course of a generation, which is roughly thirty years. Per capita income is a simple average: total income divided by the size of the population. By that measure, average income in the United States climbed from $7,787 in 1980 to $26,487 in 2010 (the latest year for which the government has data).[1] Voilà! Congratulations to us.

There is just one problem. My quick calculation is technically correct and yet totally wrong in terms of the question I set out to answer. To begin with, the figures above are not adjusted for inflation. (A per capita income of $7,787 in 1980 is equal to about $19,600 when converted to 2010 dollars.) That's a relatively quick fix. The bigger problem is that the average income in America is not equal to the income of the average American. Let's unpack that clever little phrase.

Per capita income merely takes all of the income earned in the country and divides by the number of people, which tells us absolutely nothing about who is earning how much of that income—in 1980 or in 2010. As the Occupy Wall Street folks would point out, explosive growth in the incomes of the top 1 percent can raise per capita income significantly without putting any more money in the pockets of the other 99 percent.

In other words, average income can go up without helping the average American.

As with the baseball statistic query, I have sought outside expertise on how we ought to measure the health of the American middle class. I asked two prominent labor economists, including President Obama's top economic adviser, what descriptive statistics they would use to assess the economic well-being of a typical American. Yes, you will get that answer, too, once we've taken a quick tour of descriptive statistics to give it more meaning.

From baseball to income, the most basic task when working with data is to summarize a great deal of information. There are some 330 million residents in the United States. A spreadsheet with the name and income history of every American would contain all the information we could ever want about the economic health of the country—yet it would also be so unwieldy as to tell us nothing at all. The irony is that more data can often present less clarity. So we simplify. We perform calculations that reduce a complex array of data into a handful of numbers that describe those data, just as we might encapsulate a complex, multifaceted Olympic gymnastics performance with one number: 9.8.

The good news is that these descriptive statistics give us a manageable and meaningful summary of the underlying phenomenon. That's what this chapter is about. The bad news is that any simplification invites abuse. Descriptive statistics can be like online dating profiles: technically accurate and yet pretty darn misleading.

Suppose you are at work, idly surfing the Web when you stumble across a riveting day-by-day account of Kim Kardashian's failed seventy-two-day marriage to professional basketball player Kris Humphries. You have finished reading about day seven of the marriage when your boss shows up with two enormous files of data. One file has warranty claim information for each of the 57,334 laser printers that your firm sold last year. (For each printer sold, the file documents the number of quality problems that were reported during the warranty period.) The other file has the same information for each of the 994,773 laser printers that your chief competitor sold during the same stretch. Your boss wants to know how your firm's printers compare in terms of quality with the competition.

Fortunately the computer you've been using to read about the Kardashian marriage has a basics statistics package, but where do you begin? Your instincts are probably correct: The first descriptive task is often to find some measure of the "middle" of a set of data, or what statisticians might describe as its "central tendency." What is the typical quality experience for your printers compared with those of the competition? The most basic measure of the "middle" of a distribution is the mean, or average. In this case, we want to know the average number of quality problems per printer sold for your firm and for your competitor. You would simply tally the total number of quality problems reported for all printers during the warranty period and then divide by the total number of printers sold. (Remember, the same printer can have multiple problems while under warranty.) You would do that for each firm, creating an important descriptive statistic: the average number of quality problems per printer sold.

Suppose it turns out that your competitor's printers have an average of 2.8 quality-related problems per printer during the warranty period compared with your firm's average of 9.1 reported defects. That was easy. You've just taken information on a million printers sold by two different companies and distilled it to the essence of the problem: your printers break a lot. Clearly it's time to send a short e-mail to your boss quantifying this quality gap and then get back to day eight of Kim Kardashian's marriage.

Or maybe not. I was deliberately vague earlier when I referred to the "middle" of a distribution. The mean, or average, turns out to have some problems in that regard, namely, that it is prone to distortion by "outliers," which are observations that lie farther from the center. To get your mind around this concept, imagine that ten guys are sitting on bar stools in a middle-class drinking establishment in Seattle; each of these guys earns $35,000 a year, which makes the mean annual income for the group $35,000. Bill Gates walks into the bar with a talking parrot perched on his shoulder. (The parrot has nothing to do with the example, but it kind of spices things up.) Let's assume for the sake of the example that Bill Gates has an annual income of $1 billion. When Bill sits down on the eleventh bar stool, the mean annual income for the bar patrons rises to about $91 million. Obviously none of the original ten drinkers is any richer (though it might be reasonable to expect Bill Gates to buy a round or two). If I were to describe the patrons of this bar as having an average annual income of $91 million, the statement would be both statistically correct and grossly misleading. This isn't a bar where multimillionaires hang out; it's a bar where a bunch of guys with relatively low incomes happen to be sitting next to Bill Gates and his talking parrot. The sensitivity of the mean to outliers is why we should not gauge the economic health of the American middle class by looking at per capita income. Because there has been explosive growth in incomes at the top end of the distribution—CEOs, hedge fund managers, and athletes like Derek Jeter—the average income in the United States could be heavily skewed by the megarich, making it look a lot like the bar stools with Bill Gates at the end.

For this reason, we have another statistic that also signals the "middle" of a distribution, albeit differently: the median. The median is the point that divides a distribution in half, meaning that half of the observations lie above the median and half lie below. (If there is an even number of observations, the median is the midpoint between the two middle observations.) If we return to the bar stool example, the median annual income for the ten guys originally sitting in the bar is $35,000. When Bill Gates walks in with his parrot and perches on a stool, the median annual income for the eleven of them is still $35,000. If you literally envision lining up the bar patrons on stools in ascending order of their incomes, the income of the guy sitting on the sixth stool represents the median income for the group. If Warren Buffett comes in and sits down on the twelfth stool next to Bill Gates, the median still does not change.*

---

* With twelve bar patrons, the median would be the midpoint between the income of the guy on the sixth stool and the income of the guy on the seventh stool. Since they both make $35,000, the median is $35,000. If one made $35,000 and the other made $36,000, the median for the whole group would be $35,500.

For distributions without serious outliers, the median and the mean will be similar. I've included a hypothetical summary of the quality data for the competitor's printers. In particular, I've laid out the data in what is known as a frequency distribution. The number of quality problems per printer is arrayed along the bottom; the height of each bar represents the percentages of printers sold with that number of quality problems. For example, 36 percent of the competitor's printers had two quality defects during the warranty period. Because the distribution includes all possible quality outcomes, including zero defects, the proportions must sum to 1 (or 100 percent).

Frequency Distribution of Quality Complaints for Competitor's Printers

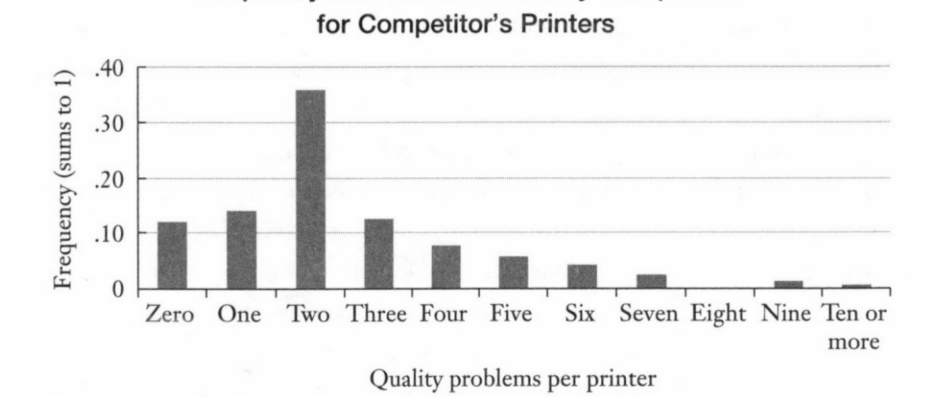

Because the distribution is nearly symmetrical, the mean and median are relatively close to one another. The distribution is slightly skewed to the right by the small number of printers with many reported quality defects. These outliers move the mean slightly rightward but have no impact on the median. Suppose that just before you dash off the quality report to your boss you decide to calculate the *median* number of quality problems for your firm's printers and the competition's. With a few keystrokes, you get the result. The median number of quality complaints for the competitor's printers is 2; the median number of quality complaints for your company's printers is 1.

Huh? Your firm's median number of quality complaints per printer is actually *lower* than your competitor's. Because the Kardashian marriage is getting monotonous, and because you are intrigued by this finding, you print a frequency distribution for your own quality problems.

Frequency Distribution of Quality Complaints at Your Company

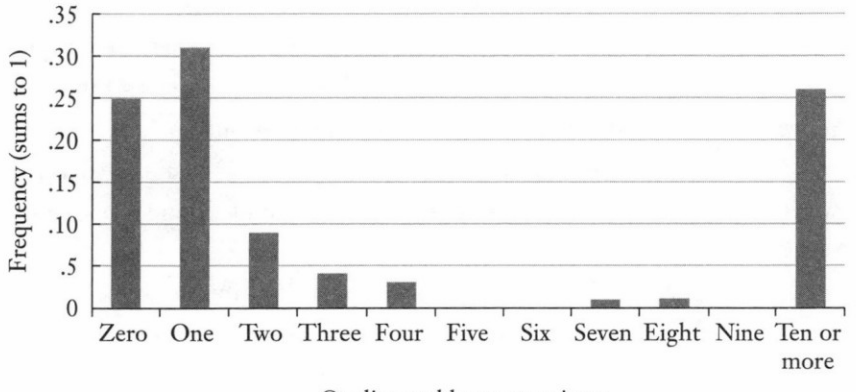

What becomes clear is that your firm does not have a uniform quality problem; you have a "lemon" problem; a small number of printers have a huge number of quality complaints. These outliers inflate the mean but not the median. More important from a production standpoint, you do not need to retool the whole manufacturing process; you need only figure out where the egregiously low-quality printers are coming from and fix that.*

Neither the median nor the mean is hard to calculate; the key is determining which measure of the "middle" is more accurate in a particular situation (a phenomenon that is easily exploited). Meanwhile, the median has some useful relatives. As we've already discussed, the median divides a distribution in half. The distribution can be further divided

* Manufacturing update: It turns out that nearly all of the defective printers were being manufactured at a plant in Kentucky where workers had stripped parts off the assembly line in order to build a bourbon distillery. Both the perpetually drunk employees and the random missing pieces on the assembly line appear to have compromised the quality of the printers being produced there.

into quarters, or quartiles. The first quartile consists of the bottom 25 percent of the observations; the second quartile consists of the next 25 percent of the observations; and so on. Or the distribution can be divided into deciles, each with 10 percent of the observations. (If your income is in the top decile of the American income distribution, you would be earning more than 90 percent of your fellow workers.) We can go even further and divide the distribution into hundredths, or percentiles. Each percentile represents 1 percent of the distribution, so that the 1st percentile represents the bottom 1 percent of the distribution and the 99th percentile represents the top 1 percent of the distribution.

The benefit of these kinds of descriptive statistics is that they describe where a particular observation lies compared with everyone else. If I tell you that your child scored in the 3rd percentile on a reading comprehension test, you should know immediately that the family should be logging more time at the library. You don't need to know anything about the test itself, or the number of questions that your child got correct. The percentile score provides a ranking of your child's score relative to that of all the other test takers. If the test was easy, then most test takers will have a high number of answers correct, but your child will have fewer correct than most of the others. If the test was extremely difficult, then all the test takers will have a low number of correct answers, but your child's score will be lower still.

Here is a good point to introduce some useful terminology. An "absolute" score, number, or figure has some intrinsic meaning. If I shoot 83 for eighteen holes of golf, that is an absolute figure. I may do that on a day that is 58 degrees, which is also an absolute figure. Absolute figures can usually be interpreted without any context or additional information. When I tell you that I shot 83, you don't need to know what other golfers shot that day in order to evaluate my performance. (The exception might be if the conditions are particularly awful, or if the course is especially difficult or easy.) If I place ninth in the golf tournament, that is a relative statistic. A "relative" value or figure has meaning only in comparison to something else, or in some broader context, such as compared with the eight golfers who shot better than I did. Most standardized tests produce results that have meaning only as a relative statistic. If I tell you that a third grader in an Illinois elementary school scored 43 out of 60 on the mathematics portion of the Illinois State Achievement Test, that absolute score doesn't have much meaning. But when I convert it to a percentile—meaning that I put that raw score into a distribution with the math scores for all other Illinois third graders—then it acquires a great deal of meaning. If 43 correct answers falls into the 83rd percentile, then this student is doing better than most of his peers statewide. If he's in the 8th percentile, then he's really struggling. In this case, the percentile (the relative score) is more meaningful than the number of correct answers (the absolute score).

Another statistic that can help us describe what might otherwise be a jumble of numbers is the standard deviation, which is a measure of how dispersed the data are from their mean. In other words, how spread out are the observations? Suppose I collected data on the weights of 250 people on an airplane headed for Boston, and I also collected the weights of a sample of 250 qualifiers for the Boston Marathon. Now assume that the mean weight for both groups is roughly the same, say 155 pounds. Anyone who has been squeezed into a row on a crowded flight, fighting for the armrest, knows that many people on a typical commercial flight weigh more than 155 pounds. But you may recall from those same unpleasant, overcrowded flights that there were lots of crying babies and poorly behaved children, all of whom have enormous lung capacity but not much mass. When it comes to calculating the average weight on the flight, the heft of the 320-pound football players on either side of your middle seat is likely offset by the tiny screaming infant across the row and the six-year-old kicking the back of your seat from the row behind.

On the basis of the descriptive tools introduced so far, the weights of the airline passengers and the marathoners are nearly identical. *But they're not.* Yes, the weights of the two groups have roughly the same "middle," but the airline passengers have far more dispersion around that midpoint, meaning that their weights are spread farther from the midpoint. My eight-year-old son might point out that the marathon runners look like they all weigh the same amount, while the airline passengers have some tiny people and some bizarrely large people. The weights of the airline passengers are "more spread out," which is an important attribute when it comes to describing the weights of these two groups. The standard deviation is the descriptive statistic that allows us to assign a single number to this dispersion around the mean. The formulas for calculating the standard deviation and the variance (another common measure of dispersion from which the standard deviation is derived) are included in an appendix at the end of the chapter. For now, let's think about why the measuring of dispersion matters.

Suppose you walk into the doctor's office. You've been feeling fatigued ever since your promotion to head of North American printer quality. Your doctor draws blood, and a few days later her assistant leaves a message on your answering machine to inform you that your HCb2 count (a fictitious blood chemical) is 134. You rush to the Internet and discover that the mean HCb2 count for a person your age is 122 (and the median is about the same). Holy crap! If you're like me, you would finally draft a will. You'd write tearful letters to your parents, spouse, children, and close friends. You might take up skydiving or try to write a novel very fast. You would send your boss a hastily composed e-mail comparing him to a certain part of the human anatomy—IN ALL CAPS.

None of these things may be necessary (and the e-mail to your boss could turn out very badly). When you call the doctor's office back to arrange for your hospice care, the physician's assistant informs you that your count is within the normal range. But how could that be? "My count is 12 points higher than average!" you yell repeatedly into the receiver.

"The standard deviation for the HCb2 count is 18," the technician informs you curtly.

What the heck does that mean?

There is natural variation in the HCb2 count, as there is with most biological phenomena (e.g., height). While the mean count for the fake chemical might be 122, plenty of healthy people have counts that are higher or lower. The danger arises only when the HCb2 count gets excessively high or low. So how do we figure out what "excessively" means in this context? As we've already noted, the standard deviation is a measure of dispersion, meaning that it reflects how tightly the observations cluster around the mean. For many typical distributions of data, a high proportion of the observations lie within one standard deviation of the mean (meaning that they are in the range from one standard deviation below the mean to one standard deviation above the mean). To illustrate with a simple example, the mean height for American adult men is 5 feet 10 inches. The standard deviation is roughly 3 inches. A high proportion of adult men are between 5 feet 7 inches and 6 feet 1 inch.

Or, to put it slightly differently, any man in this height range would not be considered abnormally short or tall. Which brings us back to your troubling HCb2 results. Yes, your count is 12 above the mean, but that's less than one standard deviation, which is the blood chemical equivalent of being about 6 feet tall—not particularly unusual. Of course, far fewer observations lie two standard deviations from the mean, and fewer still lie three or four standard deviations away. (In the case of height, an American man who is three standard deviations above average in height would be 6 feet 7 inches or taller.)

Some distributions are more dispersed than others. Hence, the standard deviation of the weights of the 250 airline passengers will be higher than the standard deviation of the weights of the 250 marathon runners. A frequency distribution with the weights of the airline passengers would literally be fatter (more spread out) than a frequency distribution of the weights of the marathon runners. Once we know the mean and standard deviation for any collection of data, we have some serious intellectual traction. For example, suppose I tell you that the mean score on the SAT math test is 500 with a standard deviation of 100. As with height, the bulk of students taking the test will be within one standard deviation of the mean, or between 400 and 600. How many students do you think score 720 or higher? Probably not very many, since that is more than two standard deviations above the mean.

In fact, we can do even better than "not very many." This is a good time to introduce one of the most important, helpful, and common distributions in statistics: the normal distribution. Data that are distributed normally are symmetrical around their mean in a bell shape that will look familiar to you.

The normal distribution describes many common phenomena.

Imagine a frequency distribution describing popcorn popping on a stove top. Some kernels start to pop early, maybe one or two pops per second; after ten or fifteen seconds, the kernels are exploding frenetically. Then gradually the number of kernels popping per second fades away at roughly the same rate at which the popping began. The heights of American men are distributed more or less normally, meaning that they are roughly symmetrical around the mean of 5 feet 10 inches. Each SAT test is specifically designed to produce a normal distribution of scores with mean 500 and standard deviation of 100. According to the *Wall Street Journal*, Americans even tend to park in a normal distribution at shopping malls; most cars park directly opposite the mall entrance—the "peak" of the normal curve—with "tails" of cars going off to the right and left of the entrance.

The beauty of the normal distribution—its Michael Jordan power, finesse, and elegance—comes from the fact that we know by definition exactly what proportion of the observations in a normal distribution lie within one standard deviation of the mean (68.2 percent), within two standard deviations of the mean (95.4 percent), within three standard deviations (99.7 percent), and so on. This may sound like trivia. In fact, it is the foundation on which much of statistics is built. We will come back to this point in much great depth later in the book.

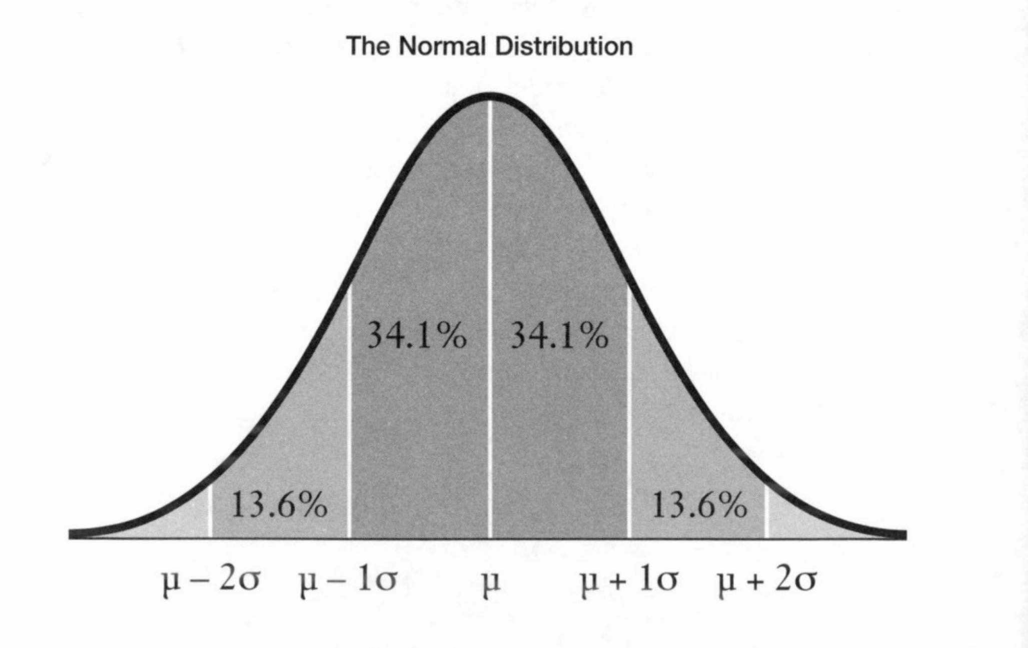

The mean is the middle line which is often represented by the Greek letter μ. The standard deviation is often represented by the Greek letter σ. Each band represents one standard deviation.

Descriptive statistics are often used to compare two figures or quantities. I'm one inch taller than my brother; today's temperature is nine degrees above the historical average for this date; and so on. Those comparisons make sense because most of us recognize the scale of the units involved. One inch does not amount to much when it comes to a person's height, so you can infer that my brother and I are roughly the same height. Conversely, nine degrees is a significant temperature deviation in just about any climate at any time of year, so nine degrees above average makes for a day that is much hotter than usual. But suppose that I told you that Granola Cereal A contains 31 milligrams more sodium than Granola Cereal B. Unless you know an awful lot about sodium (and the serving sizes for granola cereal), that statement is not going to be particularly informative. Or what if I told you that my cousin Al earned $53,000 less this year than last year? Should we be worried about Al? Or is he a hedge fund manager for whom $53,000 is a rounding error in his annual compensation?

In both the sodium and the income examples, we're missing context. The easiest way to give meaning to these relative comparisons is by using percentages. It *would* mean something if I told you that Granola Bar A has 50 percent more sodium than Granola Bar B, or that Uncle Al's income fell 47 percent last year. Measuring change as a percentage gives us some sense of scale.

You probably learned how to calculate percentages in fourth grade and will be tempted to skip the next few paragraphs. Fair enough. But first do one simple exercise for me. Assume that a department store is selling a dress for $100. The assistant manager marks down all merchandise by 25 percent. But then that assistant manager is fired for hanging out in a bar with Bill Gates,* and the new assistant manager raises all prices by

* Remarkably, this person was one of the ten people with annual incomes of $35,000 who were sitting on bar stools when Bill Gates walked in with his parrot. Go figure!

25 percent. What is the final price of the dress? If you said (or thought) $100, then you had better not skip any paragraphs.

The final price of the dress is actually $93.75. This is not merely a fun parlor trick that will win you applause and adulation at cocktail parties. Percentages are useful—but also potentially confusing or even deceptive. The formula for calculating a percentage difference (or change) is the following: (new figure – original figure)/original figure. The numerator (the part on the top of the fraction) gives us the size of the change in absolute terms; the denominator (the bottom of the fraction) is what puts this change in context by comparing it with our starting point. At first, this seems straightforward, as when the assistant store manager cuts the price of the $100 dress by 25 percent. Twenty-five percent of the original $100 price is $25; that's the discount, which takes the price down to $75. You can plug the numbers into the formula above and do some simple manipulation to get to the same place: ($100 – $75)/$100 = .25, or 25 percent.

The dress is selling for $75 when the new assistant manager demands that the price be raised 25 percent. That's where many of the people reading this paragraph probably made a mistake. The 25 percent markup is calculated as a percentage of the dress's new reduced price, which is $75. The increase will be .25($75), or $18.75, which is how the final price ends up at $93.75 (and not $100). The point is that a percentage change always gives the value of some figure *relative to something else*. Therefore, we had better understand what that something else is.

I once invested some money in a company that my college roommate started. Since it was a private venture, there were no requirements as to what information had to be provided to shareholders. A number of years went by without any information on the fate of my investment; my former roommate was fairly tight-lipped on the subject. Finally, I received a letter in the mail informing me that the firm's profits were 46 percent higher than the year before. There was no information on the size of those profits in absolute terms, meaning that I still had absolutely no idea how my investment was performing. Suppose that last year the firm earned 27 cents—essentially nothing. This year the firm earned 39 cents—also essentially nothing. Yet the company's profits grew from 27 cents to 39 cents, which is technically a 46 percent increase. Obviously the shareholder letter would have been more of a downer if it pointed out that the firm's cumulative profits over two years were less than the cost of a cup of Starbucks coffee.

To be fair to my roommate, he eventually sold the company for hundreds of millions of dollars, earning me a 100 percent return on my investment. (Since you have no idea how much I invested, you also have no idea how much money I made—which reinforces my point here very nicely!)

Let me make one additional distinction. Percentage change must not be confused with a change in percentage points. Rates are often expressed in percentages. The sales tax rate in Illinois is 6.75 percent. I pay my agent 15 percent of my book royalties. These rates are levied against some quantity, such as income in the case of the income tax rate. Obviously the rates can go up or down; less intuitively, the *changes* in the rates can be described in vastly dissimilar ways. The best example of this was a recent change in the Illinois personal income tax, which was raised from 3 percent to 5 percent. There are two ways to express this tax change, both of which are technically accurate. The Democrats, who engineered this tax increase, pointed out (correctly) that the state income tax *rate* was increased by *2 percentage points* (from 3 percent to 5 percent). The Republicans pointed out (also correctly) that the state income tax had been raised by *67 percent*. [This is a handy test of the formula from a few paragraphs back: $(5 - 3)/3 = 2/3$, which rounds up to 67 percent.]

The Democrats focused on the absolute change in the tax rate; Republicans focused on the percentage change in the tax burden. As noted, both descriptions are technically correct, though I would argue that the Republican description more accurately conveys the impact of the tax change, since what I'm going to have to pay to the government—the amount that I care about, as opposed to the way it is calculated—really has gone up by 67 percent.

Many phenomena defy perfect description with a single statistic. Suppose quarterback Aaron Rodgers throws for 365 yards but no touchdowns.

Meanwhile, Peyton Manning throws for a meager 127 yards but three touchdowns. Manning generated more points, but presumably Rodgers set up touchdowns by marching his team down the field and keeping the other team's offense off the field. Who played better? In Chapter 1, I discussed the NFL passer rating, which is the league's reasonable attempt to deal with this statistical challenge. The passer rating is an example of an index, which is a descriptive statistic made up of other descriptive statistics. Once these different measures of performance are consolidated into a single number, that statistic can be used to make comparisons, such as ranking quarterbacks on a particular day, or even over a whole career. If baseball had a similar index, then the question of the best player ever would be solved. Or would it?

The advantage of any index is that it consolidates lots of complex information into a single number. We can then rank things that otherwise defy simple comparison—anything from quarterbacks to colleges to beauty pageant contestants. In the Miss America pageant, the overall winner is a combination of five separate competitions: personal interview, swimsuit, evening wear, talent, and onstage question. (Miss Congeniality is voted on separately by the participants themselves.)

Alas, the disadvantage of any index is that it consolidates lots of complex information into a single number. There are countless ways to do that; each has the potential to produce a different outcome. Malcolm Gladwell makes this point brilliantly in a *New Yorker* piece critiquing our compelling need to rank things.[2] (He comes down particularly hard on the college rankings.) Gladwell offers the example of *Car and Driver*'s ranking of three sports cars: the Porsche Cayman, the Chevrolet Corvette, and the Lotus Evora. Using a formula that includes twenty-one different variables, *Car and Driver* ranked the Porsche number one. But Gladwell points out that "exterior styling" counts for only 4 percent of the total score in the *Car and Driver* formula, which seems ridiculously low for a sports car. If styling is given more weight in the overall ranking (25 percent), then the Lotus comes out on top.

But wait. Gladwell also points out that the sticker price of the car gets relatively little weight in the *Car and Driver* formula. If value is weighted more heavily (so that the ranking is based equally on price, exterior styling, and vehicle characteristics), the Chevy Corvette is ranked number one.

Any index is highly sensitive to the descriptive statistics that are cobbled together to build it, and to the weight given to each of those components. As a result, indices range from useful but imperfect tools to complete charades. An example of the former is the United Nations Human Development Index, or HDI. The HDI was created as a measure of economic well-being that is broader than income alone. The HDI uses income as one of its components but also includes measures of life expectancy and educational attainment. The United States ranks eleventh in the world in terms of per capita economic output (behind several oil-rich nations like Qatar, Brunei, and Kuwait) but fourth in the world in human development.[3] It's true that the HDI rankings would change slightly if the component parts of the index were reconfigured, but no reasonable change is going to make Zimbabwe zoom up the rankings past Norway. The HDI provides a handy and reasonably accurate snapshot of living standards around the globe.

Descriptive statistics give us insight into phenomena that we care about. In that spirit, we can return to the questions posed at the beginning of the chapter. Who is the best baseball player of all time? More important for the purposes of this chapter, what descriptive statistics would be most helpful in answering that question? According to Steve Moyer, president of Baseball Info Solutions, the three most valuable statistics (other than age) for evaluating any player who is not a pitcher would be the following:

1. On-base percentage (OBP), sometimes called the on-base average (OBA): Measures the proportion of the time that a player reaches base successfully, including walks (which are not counted in the batting average).
2. Slugging percentage (SLG): Measures power hitting by calculating the total bases reached per at bat. A single counts as 1,

a double is 2, a triple is 3, and a home run is 4. Thus, a batter who hit a single and a triple in five at bats would have a slugging percentage of (1 + 3)/5, or .800.

3. At bats (AB): Puts the above in context. Any mope can have impressive statistics for a game or two. A superstar compiles impressive "numbers" over thousands of plate appearances.

In Moyer's view (without hesitation, I might add), the best baseball player of all time was Babe Ruth because of his unique ability to hit and to pitch. Babe Ruth still holds the Major League career record for slugging percentage at .690.[4]

What about the economic health of the American middle class? Again, I deferred to the experts. I e-mailed Jeff Grogger (a colleague of mine at the University of Chicago) and Alan Krueger (the same Princeton economist who studied terrorists and is now serving as chair of President Obama's Council of Economic Advisers). Both gave variations on the same basic answer. To assess the economic health of America's "middle class," we should examine changes in the median wage (adjusted for inflation) over the last several decades. They also recommended examining changes to wages at the 25th and 75th percentiles (which can reasonably be interpreted as the upper and lower bounds for the middle class).

One more distinction is in order. When assessing economic health, we can examine income or wages. They are not the same thing. A wage is what we are paid for some fixed amount of labor, such as an hourly or weekly wage. Income is the sum of all payments from different sources. If workers take a second job or work more hours, their income can go up without a change in the wage. (For that matter, income can go up even if the wage is falling, provided a worker logs enough hours on the job.) However, if individuals have to work more in order to earn more, it's hard to evaluate the overall effect on their well-being. The wage is a less ambiguous measure of how Americans are being compensated for the work they do; the higher the wage, the more workers take home for every hour on the job.

Having said all that, here is a graph of American wages over the past three decades. I've also added the 90th percentile to illustrate changes in the wages for middle-class workers compared over this time frame to those workers at the top of the distribution.

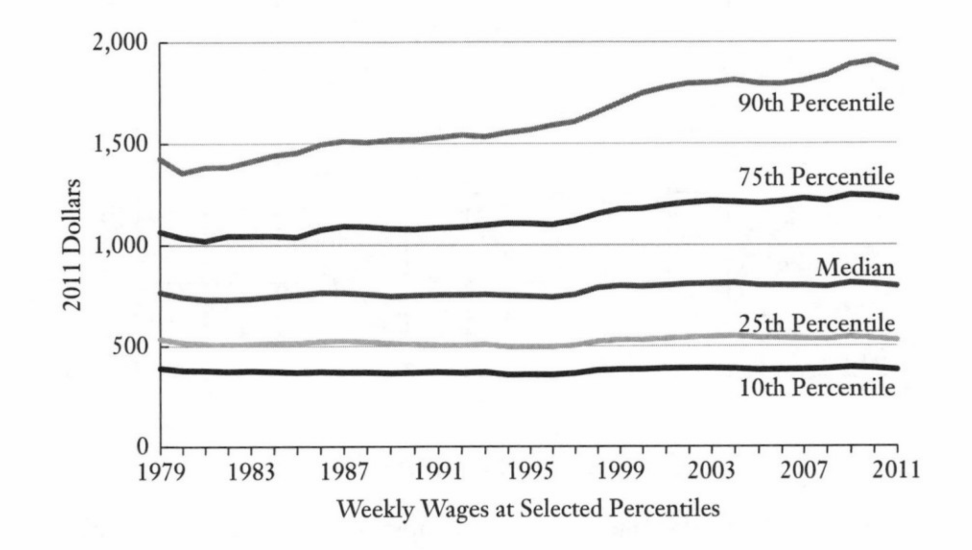

Source: "Changes in the Distribution of Workers' Hourly Wages between 1979 and 2009," Congressional Budget Office, February 16, 2011. The data for the chart can be found at http://www.cbo.gov/sites/default/files/cbofiles/ftpdocs/120xx/doc12051/02-16-wagedispersion.pdf.

A variety of conclusions can be drawn from these data. They do not present a single "right" answer with regard to the economic fortunes of the middle class. They do tell us that the typical worker, an American worker earning the median wage, has been "running in place" for nearly thirty years. Workers at the 90th percentile have done much, much better. Descriptive statistics help to frame the issue. What we do about it, if anything, is an ideological and political question.

• • •

APPENDIX TO CHAPTER 2

Data for the printer defects graphics

	Zero	One	Two	Three	Four	Five	Six	Seven	Eight	Nine	Ten or more
Frequency of competitor's defects	12	14	36	13	8	6	5	3	0	2	1
	Zero	One	Two	Three	Four	Five	Six	Seven	Eight	Nine	Ten or more
Frequency of your defects	25	31	9	4	3	0	0	1	1	0	26

Formula for variance and standard deviation

Variance and standard deviation are the most common statistical mechanisms for measuring and describing the dispersion of a distribution. The variance, which is often represented by the symbol σ^2, is calculated by determining how far the observations within a distribution lie from the mean. However, the twist is that the difference between each observation and the mean is squared; the sum of those squared terms is then divided by the number of observations.

Specifically:

For any set of n observations $x_1, x_2, x_3 \ldots x_n$ with mean μ,

$$\text{Variance} = \sigma^2 = [(x_1 - \mu)^2 + (x_2 - \mu)^2 + (x_3 - \mu)^2 + \ldots (x_n - \mu)^2]/n$$

Because the difference between each term and the mean is squared, the formula for calculating variance puts particular weight on observations that lie far from the mean, or outliers, as the following table of student heights illustrates.

Group 1	Height (μ = 70 inches)	Distance from the mean = Absolute value of $(x_n - \mu)$*	$(x_n - \mu)^2$	Group 2	Height (μ = 70 inches)	Distance from the mean = Absolute value of $(x_n - \mu)$*	$(x_n - \mu)^2$
Nick	74	4	16	Sahar	65	5	25
Elana	66	4	16	Maggie	68	2	4
Dinah	68	2	4	Faisal	69	1	1
Rebecca	69	1	1	Ted	70	0	0
Ben	73	3	9	Jeff	71	1	1
Charu	70	0	0	Narciso	75	5	25
		Total = 14	Total = 46			Total = 14	Total = 56
			Variance = 46/6 = 7.7				Variance = 56/6 = 9.3
			Standard deviation = $\sqrt{7.7} = 2.8$				Standard deviation = $\sqrt{9.3} = 3$

* Absolute value is the distance between two figures, regardless of direction, so that it is always positive. In this case, it represents the number of inches between the height of the individual and the mean.

Both groups of students have a mean height of 70 inches. The heights of students in both groups also differ from the mean by the same number of total inches: 14. By that measure of dispersion, the two distributions are identical. However, the variance for Group 2 is higher because of the weight given in the variance formula to values that lie particularly far from the mean—Sahar and Narciso in this case.

Variance is rarely used as a descriptive statistic on its own. Instead, the variance is most useful as a step toward calculating the standard deviation of a distribution, which is a more intuitive tool as a descriptive statistic.

The standard deviation for a set of observations is the square root of the variance:

For any set of n observations $x_1, x_2, x_3 \ldots x_n$ with mean μ, standard deviation = σ = square root of this whole quantity =

$$\sqrt{[(x_1 - \mu)^2 + (x_2 - \mu)^2 + (x_3 - \mu)^2 + \ldots (x_n - \mu)^2]/n}$$

CHAPTER 3

Deceptive Description

"He's got a great personality!" and other true but grossly misleading statements

To anyone who has ever contemplated dating, the phrase "he's got a great personality" usually sets off alarm bells, not because the description is necessarily wrong, but for what it may *not* reveal, such as the fact that the guy has a prison record or that his divorce is "not entirely final." We don't doubt that this guy has a great personality; we are wary that a true statement, the great personality, is being used to mask or obscure other information in a way that is seriously misleading (assuming that most of us would prefer not to date ex-felons who are still married). The statement is not a lie per se, meaning that it wouldn't get you convicted of perjury, but it still could be so inaccurate as to be untruthful.

And so it is with statistics. Although the field of statistics is rooted in mathematics, and mathematics is exact, the use of statistics to describe complex phenomena is not exact. That leaves plenty of room for shading the truth. Mark Twain famously remarked that there are three kinds of lies: lies, damned lies, and statistics.* As the last chapter explained, most phenomena that we care about can be described in multiple ways. Once

---

* Twain attributed this phrase to British prime minister Benjamin Disraeli, but there is no record of Disraeli's ever saying or writing it.

there are multiple ways of describing the same thing (e.g., "he's got a great personality" *or* "he was convicted of securities fraud"), the descriptive statistics that we choose to use (or not to use) will have a profound impact on the impression that we leave. Someone with nefarious motives can use perfectly good facts and figures to support entirely disputable or illegitimate conclusions.

We ought to begin with the crucial distinction between "precision" and "accuracy." These words are not interchangeable. Precision reflects the exactitude with which we can express something. In a description of the length of your commute, "41.6 miles" is more precise than "about 40 miles," which is more precise than "a long f——ing way." If you ask me how far it is to the nearest gas station, and I tell you that it's 1.265 miles to the east, that's a precise answer. Here is the problem: That answer may be entirely inaccurate if the gas station happens to be in the other direction. On the other hand, if I tell you, "Drive ten minutes or so until you see a hot dog stand. The gas station will be a couple hundred yards after that on the right. If you pass the Hooters, you've gone too far," my answer is less precise than "1.265 miles to the east" but significantly better because I am sending you in the direction of the gas station. Accuracy is a measure of whether a figure is broadly consistent with the truth—hence the danger of confusing precision with accuracy. If an answer is accurate, then more precision is usually better. But no amount of precision can make up for inaccuracy.

In fact, precision can mask inaccuracy by giving us a false sense of certainty, either inadvertently or quite deliberately. Joseph McCarthy, the Red-baiting senator from Wisconsin, reached the apogee of his reckless charges in 1950 when he alleged not only that the U.S. State Department was infiltrated with communists, but that he had a list of their names. During a speech in Wheeling, West Virginia, McCarthy waved in the air a piece of paper and declared, "I have here in my hand a list of 205—a list of names that were made known to the Secretary of State as being members of the Communist Party and who nevertheless are still working and shaping policy in the State Department."[1] It turns out that the paper had no names on it at all, but the specificity of the charge gave it credibility, despite the fact that it was a bald-faced lie.

I learned the important distinction between precision and accuracy in a less malicious context. For Christmas one year my wife bought me a golf range finder to calculate distances on the course from my golf ball to the hole. The device works with some kind of laser; I stand next to my ball in the fairway (or rough) and point the range finder at the flag on the green, at which point the device calculates the exact distance that I'm supposed to hit the ball. This is an improvement upon the standard yardage markers, which give distances only to the center of the green (and are therefore accurate but less precise). With my Christmas-gift range finder I was able to know that I was 147.2 yards from the hole. I expected the precision of this nifty technology to improve my golf game. Instead, it got appreciably worse.

There were two problems. First, I used the stupid device for three months before I realized that it was set to meters rather than to yards; every seemingly precise calculation (147.2) was wrong. Second, I would sometimes inadvertently aim the laser beam at the trees behind the green, rather than at the flag marking the hole, so that my "perfect" shot would go exactly the distance it was supposed to go—right over the green into the forest. The lesson for me, which applies to all statistical analysis, is that even the most precise measurements or calculations should be checked against common sense.

To take an example with more serious implications, many of the Wall Street risk management models prior to the 2008 financial crisis were quite precise. The concept of "value at risk" allowed firms to quantify with precision the amount of the firm's capital that could be lost under different scenarios. The problem was that the supersophisticated models were the equivalent of setting my range finder to meters rather than to yards. The math was complex and arcane. The answers it produced were reassuringly precise. But the assumptions about what might happen to global markets that were embedded in the models were just plain wrong, making the conclusions wholly inaccurate in ways that destabilized not only Wall Street but the entire global economy.

Even the most precise and accurate descriptive statistics can suffer from a more fundamental problem: a lack of clarity over what exactly we are trying to define, describe, or explain. Statistical arguments have much in common with bad marriages; the disputants often talk past one another. Consider an important economic question: How healthy is

American manufacturing? One often hears that American manufacturing jobs are being lost in huge numbers to China, India, and other low-wage countries. One also hears that high-tech manufacturing still thrives in the United States and that America remains one of the world's top exporters of manufactured goods. Which is it? This would appear to be a case in which sound analysis of good data could reconcile these competing narratives. Is U.S. manufacturing profitable and globally competitive, or is it shrinking in the face of intense foreign competition?

Both. The British news magazine the *Economist* reconciled the two seemingly contradictory views of American manufacturing with the following graph.

"The Rustbelt Recovery," March 10, 2011

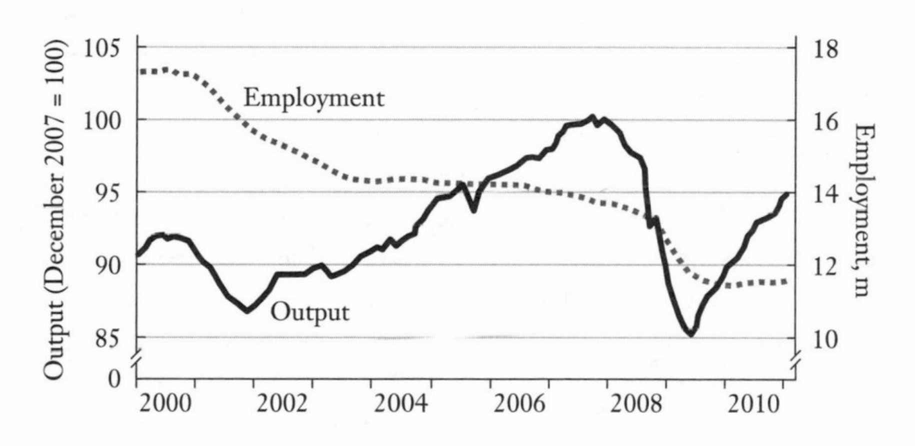

The seeming contradiction lies in how one defines the "health" of U.S. manufacturing. In terms of output—the total value of goods produced and sold—the U.S. manufacturing sector grew steadily in the 2000s, took a big hit during the Great Recession, and has since bounced back robustly. This is consistent with data from the CIA's *World Factbook* showing that the United States is the third-largest manufacturing exporter in the world, behind China and Germany. The United States remains a manufacturing powerhouse.

But the graph in the *Economist* has a second line, which is manufacturing *employment*. The number of manufacturing jobs in the United States has fallen steadily; roughly six million manufacturing jobs were lost in the last decade. Together, these two stories—rising manufacturing output and falling employment—tell the complete story. Manufacturing in the United States has grown steadily more productive, meaning that factories are producing more output with fewer workers. This is good from a global competitiveness standpoint, for it makes American products more competitive with manufactured goods from low-wage countries. (One way to compete with a firm that can pay workers $2 an hour is to create a manufacturing process so efficient that one worker earning $40 can do twenty times as much.) *But there are a lot fewer manufacturing jobs*, which is terrible news for the displaced workers who depended on those wages.

Since this is a book about statistics and not manufacturing, let's go back to the main point, which is that the "health" of U.S. manufacturing—something seemingly easy to quantify—depends on how one chooses to define health: output or employment? In this case (and many others), the most complete story comes from including both figures, as the *Economist* wisely chose to do in its graph.

Even when we agree on a single measure of success, say, student test scores, there is plenty of statistical wiggle room. See if you can reconcile the following hypothetical statements, both of which could be true:

Politician A (the challenger): "Our schools are getting worse! Sixty percent of our schools had lower test scores this year than last year."

Politician B (the incumbent): "Our schools are getting better! Eighty percent of our students had higher test scores this year than last year."

Here's a hint: The schools do not all necessarily have the same number of students. If you take another look at the seemingly contradictory statements, what you'll see is that one politician is using schools as his *unit of analysis* ("Sixty percent of our schools . . ."), and the other is using students as the unit of analysis ("Eighty percent of our students . . ."). The unit of analysis is the entity being compared or described by the statistics—school performance by one of them and student performance by the other. It's entirely possible for most of the students to be improving and most of the schools to be getting worse—if the students showing improvement happen to be in very big schools. To make this example more intuitive, let's do the same exercise by using American states:

Politician A (a populist): "Our economy is in the crapper! Thirty states had falling incomes last year."

Politician B (more of an elitist): "Our economy is showing appreciable gains: Seventy percent of Americans had rising incomes last year."

What I would infer from those statements is that the biggest states have the healthiest economies: New York, California, Texas, Illinois, and so on. The thirty states with falling average incomes are likely to be much smaller: Vermont, North Dakota, Rhode Island, and so on. Given the disparity in the size of the states, it's entirely possible that the majority of states are doing worse while the majority of Americans are doing better. The key lesson is to pay attention to the unit of analysis. Who or what is being described, and is that different from the "who" or "what" being described by someone else?

Although the examples above are hypothetical, here is a crucial statistical question that is not: Is globalization making income inequality around the planet better or worse? By one interpretation, globalization has merely exacerbated existing income inequalities; richer countries in 1980 (as measured by GDP per capita) tended to grow faster between 1980 and 2000 than poorer countries.[2] The rich countries just got richer, suggesting that trade, outsourcing, foreign investment, and the other components of "globalization" are merely tools for the developed world to extend its economic hegemony. Down with globalization! Down with globalization!

But hold on a moment. The same data can (and should) be interpreted entirely differently if one changes the unit of analysis. We don't care about poor countries; *we care about poor people*. And a high proportion of the world's poor people happen to live in China and India. Both countries are huge (with a population over a billion); each was relatively poor in 1980. Not only have China and India grown rapidly over the past several decades, but they have done so in large part because of their increased economic integration with the rest of the world. They are "rapid globalizers," as the *Economist* has described them. Given that our goal is to ameliorate human misery, it makes no sense to give China (population 1.3 billion) the same weight as Mauritius (population 1.3 million) when examining the effects of globalization on the poor.

The unit of analysis should be people, not countries. What really happened between 1980 and 2000 is a lot like my fake school example above. The bulk of the world's poor happened to live in two giant countries that grew extremely fast as they became more integrated into the global economy. The proper analysis yields an entirely different conclusion about the benefits of globalization for the world's poor. As the *Economist* points out, "If you consider people, not countries, global inequality is falling rapidly."

The telecommunications companies AT&T and Verizon have recently engaged in an advertising battle that exploits this kind of ambiguity about what is being described. Both companies provide cellular phone service. One of the primary concerns of most cell phone users is the quality of the service in places where they are likely to make or receive phone calls. Thus, a logical point of comparison between the two firms is the size and quality of their networks. While consumers just want decent cell phone service in lots of places, both AT&T and Verizon have come up with different metrics for measuring the somewhat amorphous demand for "decent cell phone service in lots of places." Verizon launched an aggressive advertising campaign touting the geographic coverage of its network; you may remember the maps of the United States that showed the large percentage of the country covered by the Verizon network compared with the relatively paltry geographic coverage of the AT&T network. The unit of analysis chosen by Verizon is geographic area covered—because the company has more of it.

AT&T countered by launching a campaign that changed the unit of analysis. Its billboards advertised that "AT&T covers 97 percent of Americans." Note the use of the word "Americans" rather than "America." AT&T focused on the fact that most people don't live in rural Montana or the Arizona desert. Since the population is not evenly distributed across the physical geography of the United States, the key to good cell service (the campaign argued implicitly) is having a network in place where callers actually live and work, not necessarily where they go camping. As someone who spends a fair bit of time in rural New Hampshire, however, my sympathies are with Verizon on this one.

Our old friends the mean and the median can also be used for nefarious ends. As you should recall from the last chapter, both the median and the mean are measures of the "middle" of a distribution, or its "central tendency." The mean is a simple average: the sum of the observations divided by the number of observations. (The mean of 3, 4, 5, 6, and 102 is 24.) The median is the midpoint of the distribution; half of the observations lie above the median and half lie below. (The median of 3, 4, 5, 6, and 102 is 5.) Now, the clever reader will see that there is a sizable difference between 24 and 5. If, for some reason, I would like to describe this group of numbers in a way that makes it look big, I will focus on the mean. If I want to make it look smaller, I will cite the median.

Now let's look at how this plays out in real life. Consider the George W. Bush tax cuts, which were touted by the Bush administration as something good for most American families. While pushing the plan, the administration pointed out that 92 million Americans would receive an average tax reduction of over $1,000 ($1,083 to be precise). But was that summary of the tax cut accurate? According to the *New York Times*, "The data don't lie, but some of them are mum."

Would 92 million Americans be getting a tax cut? Yes.

Would most of those people be getting a tax cut of around $1,000? No. The median tax cut was less than $100.

A relatively small number of extremely wealthy individuals were eligible for very large tax cuts; these big numbers skew the mean, making the average tax cut look bigger than what most Americans would likely receive. The median is not sensitive to outliers, and, in this case, is probably a more accurate description of how the tax cuts affected the typical household.

Of course, the median can also do its share of dissembling *because it is not sensitive to outliers*. Suppose that you have a potentially fatal illness. The good news is that a new drug has been developed that might be effective. The drawback is that it's extremely expensive and has many unpleasant side effects. "But does it work?" you ask. The doctor informs you that the new drug increases the median life expectancy among patients with your disease by two weeks. That is hardly encouraging news; the drug may not be worth the cost and unpleasantness. Your insurance company refuses to pay for the treatment; it has a pretty good case on the basis of the median life expectancy figures.

Yet the median may be a horribly misleading statistic in this case. Suppose that many patients do not respond to the new treatment but that some large number of patients, say 30 or 40 percent, are cured entirely. This success would not show up in the median (though the mean life expectancy of those taking the drug would look very impressive). In this case, the outliers—those who take the drug and live for a long time—would be highly relevant to your decision. And it is not merely a hypothetical case. Evolutionary biologist Stephen Jay Gould was diagnosed with a form of cancer that had a median survival time of eight months; he died of a different and unrelated kind of cancer twenty years later.[3] Gould subsequently wrote a famous article called "The Median Isn't the Message," in which he argued that his scientific knowledge of statistics saved him from the erroneous conclusion that he would necessarily be dead in eight months. The definition of the median tells us that half the patients will live at least eight months—and possibly much, much longer than that. The mortality distribution is "right-skewed," which is more than a technicality if you happen to have the disease.[4]

In this example, the defining characteristic of the median—that it does not weight observations on the basis of *how far* they lie from the midpoint, only on whether they lie above or below—turns out to be its weakness. In contrast, the mean *is* affected by dispersion. From the standpoint of accuracy, the median versus mean question revolves around whether the outliers in a distribution distort what is being described or are instead an important part of the message. (Once again, judgment trumps math.) Of course, nothing says that you must choose the median or the mean. Any comprehensive statistical analysis would likely present both. When just the median or the mean appears, it may be for the sake of brevity—or it may be because someone is seeking to "persuade" with statistics.

Those of a certain age may remember the following exchange (as I recollect it) between the characters played by Chevy Chase and Ted Knight in the movie *Caddyshack*. The two men meet in the locker room after both have just come off the golf course:

> TED KNIGHT: What did you shoot?
> CHEVY CHASE: Oh, I don't keep score.
> TED KNIGHT: Then how do you compare yourself to other golfers?
> CHEVY CHASE: By height.

I'm not going to try to explain why this is funny. I will say that a great many statistical shenanigans arise from "apples and oranges" comparisons. Suppose you are trying to compare the price of a hotel room in London with the price of a hotel room in Paris. You send your six-year-old to the computer to do some Internet research, since she is much faster and better at it than you are. Your child reports back that hotel rooms in Paris are more expensive, around 180 a night; a comparable room in London is 150 a night.

You would likely explain to your child the difference between pounds and euros, and then send her back to the computer to find the exchange rate between the two currencies so that you could make a meaningful comparison. (This example is loosely rooted in truth; after I paid 100 rupees for a pot of tea in India, my daughter wanted to know why everything in India was so expensive.) Obviously the numbers on currency from different countries mean nothing until we convert them into comparable units. What is the exchange rate between the pound and the euro, or, in the case of India, between the dollar and the rupee?

This seems like a painfully obvious lesson—yet one that is routinely ignored, particularly by politicians and Hollywood studios. These folks clearly recognize the difference between euros and pounds; instead, they overlook a more subtle example of apples and oranges: inflation. A dollar today is not the same as a dollar sixty years ago; it buys much less. Because of inflation, something that cost $1 in 1950 would cost $9.37 in 2011. As a result, any monetary comparison between 1950 and 2011 without adjusting for changes in the value of the dollar would be less accurate than comparing figures in euros and pounds—*since the euro and the pound are closer to each other in value than a 1950 dollar is to a 2011 dollar.*

This is such an important phenomenon that economists have terms to denote whether figures have been adjusted for inflation or not. *Nominal* figures are not adjusted for inflation. A comparison of the nominal cost of a government program in 1970 to the nominal cost of the same program in 2011 merely compares the size of the checks that the Treasury wrote in those two years—without any recognition that a dollar in 1970 bought more stuff than a dollar in 2011. If we spent $10 million on a program in 1970 to provide war veterans with housing assistance and $40 million on the same program in 2011, *the federal commitment to that program has actually gone down*. Yes, spending has gone up in nominal terms, but that does not reflect the changing value of the dollars being spent. One 1970 dollar is equal to $5.83 in 2011; the government would need to spend $58.3 million on veterans' housing benefits in 2011 to provide support comparable to the $10 million it was spending in 1970.

Real figures, on the other hand, are adjusted for inflation. The most commonly accepted methodology is to convert all of the figures into a single unit, such as 2011 dollars, to make an "apples and apples" comparison. Many websites, including that of the U.S. Bureau of Labor Statistics, have simple inflation calculators that will compare the value of a dollar at different points in time.* For a real (yes, a pun) example of how statistics can look different when adjusted for inflation, check out the following graph of the U.S. federal minimum wage, which plots both the nominal value of the minimum wage and its real purchasing power in 2010 dollars.

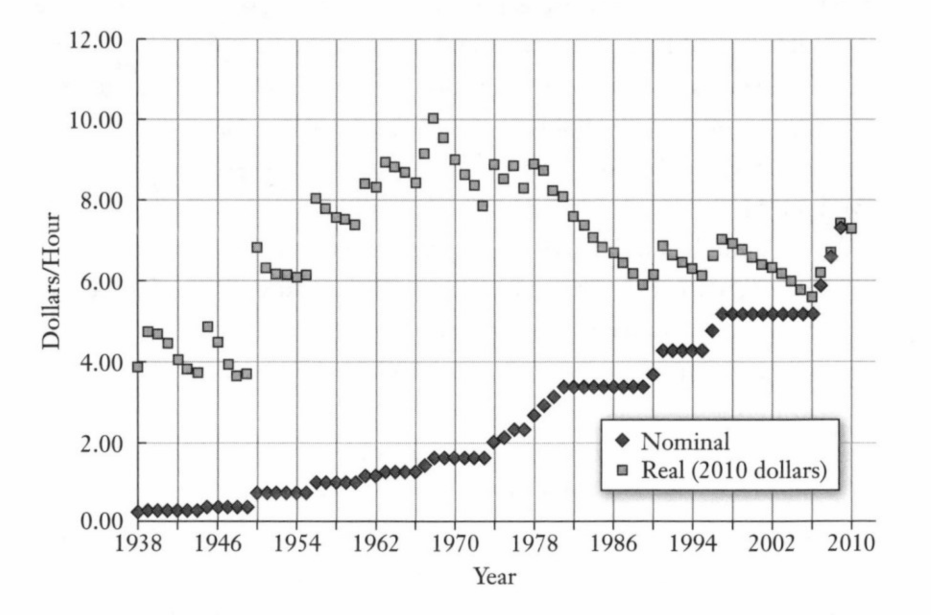

Source: http://oregonstate.edu/instruct/anth484/minwage.html.

* Available at http://www.bls.gov/data/inflation_calculator.htm.

The federal minimum wage—the number posted on the bulletin board in some remote corner of your office—is set by Congress. This wage, currently $7.25, is a nominal figure. Your boss does not have to ensure that $7.25 buys as much as it did two years ago; he just has to make sure that you get a minimum of $7.25 for every hour of work that you do. It's all about the number on the check, not what that number can buy.

Yet inflation erodes the purchasing power of the minimum wage over time (and every other nominal wage, which is why unions typically negotiate "cost of living adjustments"). If prices rise faster than Congress raises the minimum wage, the real value of that minimum hourly payment will fall. Supporters of a minimum wage should care about the real value of that wage, since the whole point of the law is to guarantee low-wage workers some minimum level of consumption for an hour of work, not to give them a check with a big number on it that buys less than it used to. (If that were the case, then we could just pay low-wage workers in rupees.)

Hollywood studios may be the most egregiously oblivious to the distortions caused by inflation when comparing figures at different points in time—and deliberately so. What were the top five highest-grossing films (domestic) of all time as of 2011?[5]

1. *Avatar* (2009)
2. *Titanic* (1997)
3. *The Dark Knight* (2008)
4. *Star Wars Episode IV* (1977)
5. *Shrek 2* (2004)

Now you may feel that list looks a little suspect. These were successful films—but *Shrek 2*? Was that really a greater commercial success than *Gone with the Wind*? *The Godfather*? *Jaws*? No, no, and no. Hollywood likes to make each blockbuster look bigger and more successful than the last. One way to do that would be to quote box office receipts in Indian rupees, which would inspire headlines such as the following: "Harry Potter Breaks Box Office Record with Weekend Receipts of 1.3 Trillion!" But even the most dim-witted moviegoers would be suspicious of figures that are large only because they are quoted in a currency with relatively little purchasing power. Instead, Hollywood studios (and the journalists who report on them) merely use nominal figures, which makes recent movies look successful largely because ticket prices are higher now than they were ten, twenty, or fifty years ago. (When *Gone with the Wind* came out in 1939, a ticket cost somewhere in the range of $.50.) The most accurate way to compare commercial success over time would be to adjust ticket receipts for inflation. Earning $100 million in 1939 is a lot more impressive than earning $500 million in 2011. So what are the top grossing films in the U.S. of all time, *adjusted for inflation*?[6]

1. *Gone with the Wind* (1939)
2. *Star Wars Episode IV* (1977)
3. *The Sound of Music* (1965)
4. *E.T.* (1982)
5. *The Ten Commandments* (1956)

In real terms, *Avatar* falls to number 14; *Shrek 2* falls all the way to 31st.

Even comparing apples and apples leaves plenty of room for shenanigans. As discussed in the last chapter, one important role of statistics is to describe changes in quantities over time. Are taxes going up? How many cheeseburgers are we selling compared with last year? By how much have we reduced the arsenic in our drinking water? We often use percentages to express these changes because they give us a sense of scale and context. We understand what it means to reduce the amount of arsenic in the drinking water by 22 percent, whereas few of us would know whether reducing arsenic by one microgram (the absolute reduction) would be a significant change or not. Percentages don't lie—but they can exaggerate. One way to make growth look explosive is to use percentage change to describe some change relative to a very low starting point. I live in Cook County, Illinois. I was shocked one day to learn that the portion of my taxes supporting the Suburban Cook County Tuberculosis Sanitarium District was slated to rise by 527 percent! However, I called off my massive antitax rally (which was really still in the planning phase) when I learned that this change would cost me less than a good turkey sandwich. The Tuberculosis Sanitarium District deals with roughly a hundred cases a year; it is not a large or expensive organization. The *Chicago Sun-Times* pointed out that for the typical homeowner, the tax bill would go from $1.15 to $6.[7] Researchers will sometimes qualify a growth figure by pointing out that it is "from a low base," meaning that any increase is going to look large by comparison.

Obviously the flip side is true. A small percentage of an enormous sum can be a big number. Suppose the secretary of defense reports that defense spending will grow only 4 percent this year. Great news! Not really, given that the Defense Department budget is nearly $700 billion. Four percent of $700 billion is $28 billion, which can buy a lot of turkey sandwiches. In fact, that seemingly paltry 4 percent increase in the defense budget is ***more than the entire NASA budget and about the same as the budgets of the Labor and Treasury Departments combined.***

In a similar vein, your kindhearted boss might point out that as a matter of fairness, every employee will be getting the same raise this year, 10 percent. What a magnanimous gesture—except that if your boss makes $1 million and you make $50,000, his raise will be $100,000 and yours will be $5,000. The statement "everyone will get the same 10 percent raise this year" just sounds so much better than "my raise will be twenty times bigger than yours." Both are true in this case.

Any comparison of a quantity changing over time must have a start point and an end point. One can sometimes manipulate those points in ways that affect the message. I once had a professor who liked to speak about his "Republican slides" and his "Democratic slides." He was referring to data on defense spending, and what he meant was that he could organize the same data in different ways in order to please either Democratic or Republican audiences. For his Republican audiences, he would offer the following slide with data on increases in defense spending under Ronald Reagan. Clearly Reagan helped restore our commitment to defense and security, which in turn helped to win the Cold War. No one can look at these numbers and not appreciate the steely determination of Ronald Reagan to face down the Soviets.

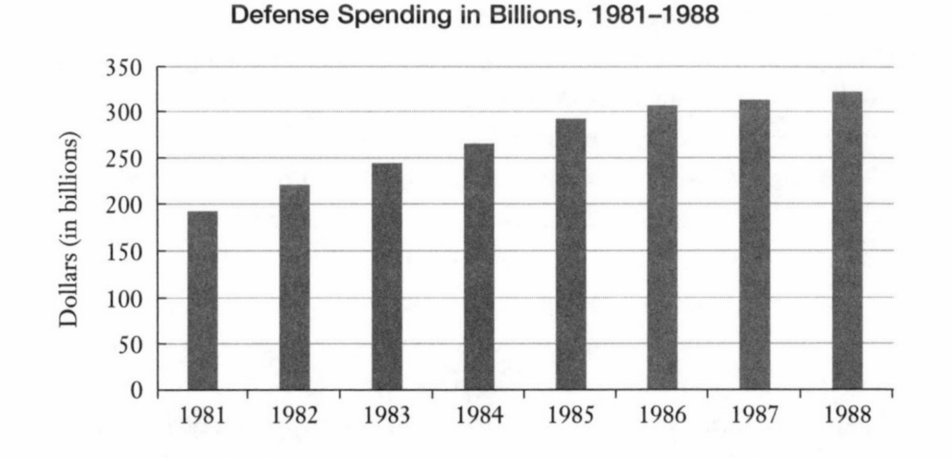

For the Democrats, my former professor merely used the same (nominal) data, but a longer time frame. For this group, he pointed out that Jimmy Carter deserves credit for beginning the defense buildup. As the following "Democratic" slide shows, the defense spending increases from 1977 to 1980 show the same basic trend as the increases during the Reagan presidency. Thank goodness that Jimmy Carter—a graduate of Annapolis and a former naval officer—began the process of making America strong again!

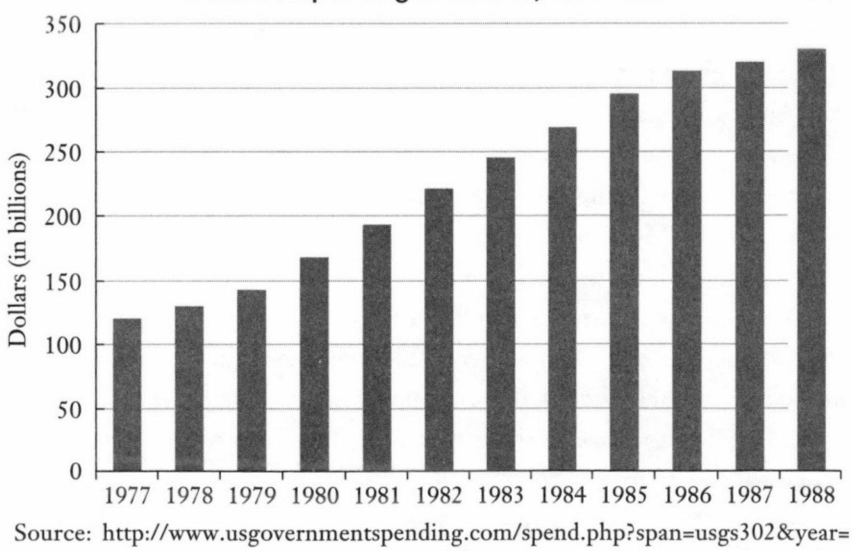

Source: http://www.usgovernmentspending.com/spend.php?span=usgs302&year=1988&view=1&expand=30&expandC=&units=b&fy=fy12&local=s&state=US&pie=#usgs302.

While the main point of statistics is to present a meaningful picture of things we care about, in many cases we also hope to act on these numbers. NFL teams want a simple measure of quarterback quality so that they can find and draft talented players out of college. Firms measure the performance of their employees so that they can promote those who are valuable and fire those who are not. There is a common business aphorism: "You can't manage what you can't measure." True. *But you had better be darn sure that what you are measuring is really what you are trying to manage.*

Consider school quality. This is a crucial thing to measure, since we would like to reward and emulate "good" schools while sanctioning or fixing "bad" schools. (And within each school, we have the similar challenge of measuring teacher quality, for the same basic reason.) The most common measure of quality for both schools and teachers is test scores. If students are achieving impressive scores on a well-conceived standardized test, then presumably the teacher and school are doing a fine job. Conversely, bad test scores are a clear signal that lots of people should be fired, sooner rather than later. These statistics can take us a long way toward fixing our public education system, right?

Wrong. Any evaluation of teachers or schools that is based solely on test scores will present a dangerously inaccurate picture. Students who walk through the front door of different schools have vastly different backgrounds and abilities. We know, for example, that the education and income of a student's parents have a significant impact on achievement, regardless of what school he or she attends. The statistic that we're missing in this case happens to be the only one that matters for our purposes: How much of a student's performance, good or bad, can be attributed to what happens inside the school (or inside a particular classroom)?

Students who live in affluent, highly educated communities are going to test well from the moment their parents drop them off at school on the first day of kindergarten. The flip side is also true. There are schools with extremely disadvantaged populations in which teachers may be doing a remarkable job but the student test scores will still be low—albeit not nearly as low as they would have been if the teachers had not been doing a good job. What we need is some measure of "value-added" at the school level, or even at the classroom level. We don't want to know the absolute level of student achievement; we want to know how much that student achievement has been affected by the educational factors we are trying to evaluate.

At first glance, this seems an easy task, as we can simply give students a pretest and a posttest. If we know student test scores when they enter a particular school or classroom, then we can measure their performance at the end and attribute the difference to whatever happened in that school or classroom.

Alas, wrong again. Students with different abilities or backgrounds may also learn *at different rates*. Some students will grasp the material faster than others for reasons that have nothing to do with the quality of the teaching. So if students in Affluent School A and Poor School B both start algebra at the same time and level, the explanation for the fact that students at Affluent School A test better in algebra a year later may be that the teachers are better, or it may be that the students were capable of learning faster—or both. Researchers are working to develop statistical techniques that measure instructional quality in ways that account appropriately for different student backgrounds and abilities. In the meantime, our attempts to identify the "best" schools can be ridiculously misleading.

Every fall, several Chicago newspapers and magazines publish a ranking of the "best" high schools in the region, usually on the basis of state test score data. Here is the part that is laugh-out-loud funny from a statistical standpoint: Several of the high schools consistently at the top of the rankings are selective enrollment schools, meaning that students must apply to get in, and only a small proportion of those students are accepted. One of the most important admissions criteria is standardized test scores. So let's summarize: (1) these schools are being recognized as "excellent" for having students with high test scores; (2) to get into such a school, one must have high test scores. This is the logical equivalent of giving an award to the basketball team for doing such an excellent job of producing tall students.

Even if you have a solid indicator of what you are trying to measure and manage, the challenges are not over. The good news is that "managing by statistics" can change the underlying behavior of the person or institution being managed for the better. If you can measure the proportion of defective products coming off an assembly line, and if those defects are a function of things happening at the plant, then some kind of bonus for workers that is tied to a reduction in defective products would presumably change behavior in the right kinds of ways. Each of us responds to incentives (even if it is just praise or a better parking spot). Statistics measure the outcomes that matter; incentives give us a reason to improve those outcomes.

Or, in some cases, just to make the statistics look better. That's the bad news.

If school administrators are evaluated—and perhaps even compensated—on the basis of the high school graduation rate for students in a particular school district, they will focus their efforts on boosting the number of students who graduate. Of course, they may also devote some effort to improving the graduation rate, which is not necessarily the same thing. For example, students who leave school before graduation can be classified as "moving away" rather than dropping out. This is not merely a hypothetical example; it is a charge that was leveled against former secretary of education Rod Paige during his tenure as the Houston school superintendent. Paige was hired by President George W. Bush to be U.S. secretary of education because of his remarkable success in Houston in reducing the dropout rate and boosting test scores.

If you're keeping track of the little business aphorisms I keep tossing your way, here is another one: "It's never a good day when *60 Minutes* shows up at your door." Dan Rather and the *60 Minutes II* crew made a trip to Houston and found that the manipulation of statistics was far more impressive than the educational improvement.[8] High schools routinely classified students who quit high school as transferring to another school, returning to their native country, or leaving to pursue a General Equivalency Diploma (GED)—none of which count as dropping out in the official statistics. Houston reported a citywide dropout rate of 1.5 percent in the year that was examined; *60 Minutes* calculated that the true dropout rate was between 25 and 50 percent.

The statistical chicanery with test scores was every bit as impressive. One way to improve test scores (in Houston or anywhere else) is to improve the quality of education so that students learn more and test better. This is a good thing. Another (less virtuous) way to improve test scores is to prevent the worst students from taking the test. If the scores of the lowest-performing students are eliminated, the average test score for the school or district will go up, even if all the rest of the students show no improvement at all. In Texas, the statewide achievement test is given in tenth grade. There was evidence that Houston schools were trying to keep the weakest students from reaching tenth grade. In one particularly egregious example, a student spent three years in ninth grade and then was promoted straight to eleventh grade—a deviously clever way of keeping a weak student from taking a tenth-grade benchmark exam without forcing him to drop out (which would have showed up on a different statistic).

It's not clear that Rod Paige was complicit in this statistical trickery during his tenure as Houston superintendent; however, he did implement a rigorous accountability program that gave cash bonuses to principals who met their dropout and test score goals and that fired or demoted principals who failed to meet their targets. Principals definitely responded to the incentives; that's the larger lesson. But you had better be darn certain that the folks being evaluated can't make themselves look better (statistically) in ways that are not consistent with the goal at hand.

The state of New York learned this the hard way. The state introduced "scorecards" that evaluate the mortality rates for the patients of cardiologists performing coronary angioplasty, a common treatment for heart disease.[9] This seems like a perfectly reasonable and helpful use of descriptive statistics. The proportion of a cardiologist's patients who die in surgery is an important thing to know, and it makes sense for the government to collect and promulgate such data since individual consumers would not otherwise have access to it. So is this a good policy? Yes, other than the fact that it probably ended up killing people.

Cardiologists obviously care about their "scorecard." However, the easiest way for a surgeon to improve his mortality rate is *not* by killing fewer people; presumably most doctors are already trying very hard to keep their patients alive. The easiest way for a doctor to improve his mortality rate is by refusing to operate on the sickest patients. According to a survey conducted by the School of Medicine and Dentistry at the University of Rochester, the scorecard, which ostensibly serves patients, can also work to their detriment: 83 percent of the cardiologists surveyed said that, because of the public mortality statistics, some patients who might benefit from angioplasty might not receive the procedure; 79 percent of the doctors said that some of their personal medical decisions had been influenced by the knowledge that mortality data are collected and made public. The sad paradox of this seemingly helpful descriptive statistic is that cardiologists responded rationally by withholding care from the patients who needed it most.

A statistical index has all the potential pitfalls of any descriptive statistic—plus the distortions introduced by combining multiple indicators into a single number. By definition, any index is going to be sensitive to how it is constructed; it will be affected both by what measures go into the index and by how each of those measures is weighted. For example, why does the NFL passer rating not include any measure of third down completions? And for the Human Development Index, how should a country's literacy rate be weighted in the index relative to per capita income? In the end, the important question is whether the simplicity and ease of use introduced by collapsing many indicators into a single number outweighs the inherent inaccuracy of the process. Sometimes that answer may be no, which brings us back (as promised) to the *U.S. News & World Report* (*USNWR*) college rankings.

The *USNWR* rankings use sixteen indicators to score and rank America's colleges, universities, and professional schools. In 2010, for example, the ranking of national universities and liberal arts colleges used "student selectivity" as 15 percent of the index; student selectivity is in turn calculated on the basis of a school's acceptance rate, the proportion of the entering students who were in the top 10 percent of their high school class, and the average SAT and ACT scores of entering students. The benefit of the *USNWR* rankings is that they provide lots of information about thousands of schools in a simple and accessible way. Even the critics concede that much of the information collected on America's colleges and universities is valuable. Prospective students should know an institution's graduation rate and the average class size.

Of course, providing meaningful information is an enterprise entirely different from that of collapsing all of that information into a single ranking that purports to be authoritative. To critics, the rankings are sloppily constructed, misleading, and detrimental to the long-term interests of students. "One concern is simply about its being a list that claims to rank institutions in numerical order, which is a level of precision that those data just don't support," says Michael McPherson, the former president of Macalester College in Minnesota.[10] Why should alumni giving count for 5 percent of a school's score? And if it's important, why does it not count for ten percent?

According to *U.S. News & World Report*, "Each indicator is assigned a weight (expressed as a percentage) based on our judgments about which measures of quality matter most."[11] Judgment is one thing; arbitrariness is another. The most heavily weighted variable in the ranking of national universities and colleges is "academic reputation." This reputation is determined on the basis of a "peer assessment survey" filled out by administrators at other colleges and universities and from a survey of high school guidance counselors. In his general critique of rankings, Malcolm Gladwell offers a scathing (though humorous) indictment of the peer assessment methodology. He cites a questionnaire sent out by a former chief justice of the Michigan Supreme Court to roughly one hundred lawyers asking them to rank ten law schools in order of quality. Penn State's was one of the law schools on the list; the lawyers ranked it near the middle. *At the time, Penn State did not have a law school.*[12]

For all the data collected by *USNWR*, it's not obvious that the rankings measure what prospective students ought to care about: How much learning is going on at any given institution? Football fans may quibble about the composition of the passer index, but no one can deny that its component parts—completions, yardage, touchdowns, and interceptions—are an important part of a quarterback's overall performance. That is not necessarily the case with the *USNWR* criteria, most of which focus on inputs (e.g., what kind of students are admitted, how much faculty are paid, the percentage of faculty who are full-time) rather than educational outputs. Two notable exceptions are the freshman retention rate and the graduation rate, but even those indicators do not measure learning. As

Michael McPherson points out, "We don't really learn anything from U.S. News about whether the education they got during those four years actually improved their talents or enriched their knowledge."

All of this would still be a harmless exercise, but for the fact that it appears to encourage behavior that is not necessarily good for students or higher education. For example, one statistic used to calculate the rankings is financial resources per student; the problem is that there is no corresponding measure of how well that money is being spent. An institution that spends less money to better effect (and therefore can charge lower tuition) is punished in the ranking process. Colleges and universities also have an incentive to encourage large numbers of students to apply, including those with no realistic hope of getting in, because it makes the school appear more selective. This is a waste of resources for the schools soliciting bogus applications and for students who end up applying with no meaningful chance of being accepted.

Since we are about to move on to a chapter on probability, I will bet that the *U.S. News & World Report* rankings are not going away anytime soon. As Leon Botstein, president of Bard College, has pointed out, "People love easy answers. What is the best place? Number 1."[13]

The overall lesson of this chapter is that statistical malfeasance has very little to do with bad math. If anything, impressive calculations can obscure nefarious motives. The fact that you've calculated the mean correctly will not alter the fact that the median is a more accurate indicator. Judgment and integrity turn out to be surprisingly important. A detailed knowledge of statistics does not deter wrongdoing any more than a detailed knowledge of the law averts criminal behavior. With both statistics and crime, the bad guys often know exactly what they're doing!

CHAPTER 4

Correlation

How does Netflix know what movies I like?

Netflix insists that I'll like the film *Bhutto*, a documentary that offers an "in-depth and at times incendiary look at the life and tragic death of former Pakistani prime minister Benazir Bhutto." I probably will like the film *Bhutto*. (I've added it to my queue.) The Netflix recommendations that I've watched in the past have been terrific. And when a film is recommended that I've already seen, it's typically one I've really enjoyed.

How does Netflix do that? Is there some massive team of interns at corporate headquarters who have used a combination of Google and interviews with my family and friends to determine that I might like a documentary about a former Pakistani prime minister? Of course not. Netflix has merely mastered some very sophisticated statistics. *Netflix doesn't know me*. But it does know what films I've liked in the past (because I've rated them). Using that information, along with ratings from other customers and a powerful computer, Netflix can make shockingly accurate predictions about my tastes.

I'll come back to the specific Netflix algorithm for making these picks; for now, the important point is that it's all based on correlation. Netflix recommends movies that are similar to other films that I've liked; it also recommends films that have been highly rated by other customers whose ratings are similar to mine. *Bhutto* was recommended because of my five-star ratings for two other documentaries, *Enron: The Smartest Guys in the Room* and *Fog of War*.

Correlation measures the degree to which two phenomena are related to one another. For example, there is a correlation between summer temperatures and ice cream sales. When one goes up, so does the other. Two variables are positively correlated if a change in one is associated with a change in the other in the same direction, such as the relationship between height and weight. Taller people weigh more (on average); shorter people weigh less. A correlation is negative if a positive change in one variable is associated with a negative change in the other, such as the relationship between exercise and weight.

The tricky thing about these kinds of associations is that not every observation fits the pattern. Sometimes short people weigh more than tall people. Sometimes people who don't exercise are skinnier than people who exercise all the time. Still, there is a meaningful relationship between height and weight, and between exercise and weight.

If we were to do a scatter plot of the heights and weights of a random sample of American adults, we would expect to see something like the following:

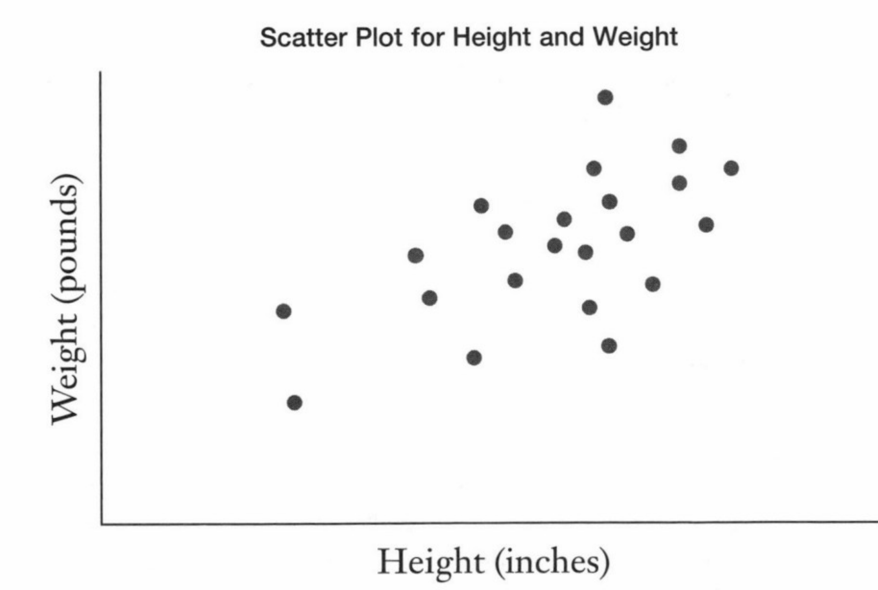

If we were to create a scatter plot of the association between exercise (as measured by minutes of intensive exercise per week) and weight, we would expect a negative correlation, with those who exercise more tending to weigh less. But a pattern consisting of dots scattered across the page is a somewhat unwieldy tool. (If Netflix tried to make film recommendations for me by plotting the ratings for thousands of films by millions of customers, the results would bury the headquarters in scatter plots.) Instead, the power of correlation as a statistical tool is that we can encapsulate an association between two variables in a single descriptive statistic: the correlation coefficient.

The correlation coefficient has two fabulously attractive characteristics. First, for math reasons that have been relegated to the appendix, it is a single number ranging from –1 to 1. A correlation of 1, often described as perfect correlation, means that every change in one variable is associated with an equivalent change in the other variable in the same direction.

A correlation of –1, or perfect negative correlation, means that every change in one variable is associated with an equivalent change in the other variable in the opposite direction.

The closer the correlation is to 1 or –1, the stronger the association. A correlation of 0 (or close to it) means that the variables have no meaningful association with one another, such as the relationship between shoe size and SAT scores.

The second attractive feature of the correlation coefficient is that it has no units attached to it. We can calculate the correlation between height and weight—even though height is measured in inches and weight is measured in pounds. We can even calculate the correlation between the number of televisions high school students have in their homes and their SAT scores, which I assure you will be positive. (More on that relationship in a moment.) The correlation coefficient does a seemingly miraculous thing: It collapses a complex mess of data measured in different units (like our scatter plots of height and weight) into a single, elegant descriptive statistic.

How?

As usual, I've put the most common formula for calculating the correlation coefficient in the appendix at the end of the chapter. This is not a statistic that you are going to be calculating by hand. (After you've entered the data, a basic software package like Microsoft Excel will calculate the correlation between two variables.) Still, the intuition is not that difficult. The formula for calculating the correlation coefficient does the following:

1. Calculates the mean and standard deviation for both variables. If we stick with the height and weight example, we would then know the mean height for people in the sample, the mean weight for people in the sample, and the standard deviation for both height and weight.
2. Converts all the data so that each observation is represented by its distance (in standard deviations) from the mean. Stick with me; it's not that complicated. Suppose that the mean height in the sample is 66 inches (with a standard deviation of 5 inches) and that the mean weight is 177 pounds (with a standard deviation of 10 pounds). Now suppose that you are 72 inches tall and weigh 168 pounds. We can also say that you your height is 1.2 standard deviations above the mean in height [(72 – 66)/5)] and .9 standard deviations below the mean in weight, or –0.9 for purposes of the formula [(168 – 177)/10]. *Yes, it's unusual for someone to be above the mean in height and below the mean in weight, but since you've paid good money for this book, I figured I should at least make you tall and thin.* Notice that your height and weight, formerly in inches and pounds, have been reduced to 1.2 and –0.9. This is what makes the units go away.
3. Here I'll wave my hands and let the computer do the work. The formula then calculates the relationship between height and weight across all the individuals in the sample as measured by standard units. When individuals in the sample are tall, say, 1.5 or 2 standard deviations above the mean, what do their weights tend to be *as measured in standard deviations from the mean for weight*? And when individuals are near to the mean in terms of height, what are their weights as measured in standard units?

If the distance from the mean for one variable tends to be broadly consistent with distance from the mean for the other variable (e.g., people who are far from the mean for height in either direction tend also to be far from the mean in the same direction for weight), then we would expect a strong positive correlation.

If distance from the mean for one variable tends to correspond to a similar distance from the mean for the second variable *in the other direction* (e.g., people who are far above the mean in terms of exercise tend to be far below the mean in terms of weight), then we would expect a strong negative correlation.

If two variables do not tend to deviate from the mean in any meaningful pattern (e.g., shoe size and exercise) then we would expect little or no correlation.

You suffered mightily in that section; we'll get back to film rentals soon. Before we return to Netflix, however, let's reflect on another aspect of life where correlation matters: the SAT. Yes, that SAT. The SAT Reasoning Test, formerly known as the Scholastic Aptitude Test, is a standardized exam made up of three sections: math, reading, and writing. You probably took the SAT, or will soon. You probably did not reflect deeply on *why* you had to take the SAT. The purpose of the test is to measure academic ability and predict college performance. Of course, one might reasonably ask (particularly those who don't like standardized tests): Isn't that what high school is for? Why is a four-hour test so important when college admissions officers have access to *four years* of high school grades?

The answer to those questions is lurking back in Chapters 1 and 2. High school grades are an imperfect descriptive statistic. A student who gets mediocre grades while taking a tough schedule of math and science classes may have more academic ability and potential than a student at the same school with better grades in less challenging classes. Obviously there are even larger potential discrepancies across schools. According to the College Board, which produces and administers the SAT, the test was created to "democratize access to college for all students." Fair enough. The SAT offers a standardized measure of ability that can be compared easily across all students applying to college. *But is it a good measure of ability?* If we want a metric that can be compared easily across students, we could also have all high school seniors run the 100 yard dash, which is cheaper and easier than administering the SAT. The problem, of course, is that performance in the 100 yard dash is uncorrelated with college performance. It's easy to get the data; they just won't tell us anything meaningful.

So how well does the SAT fare in this regard? Sadly for future generations of high school students, the SAT does a reasonably good job of predicting first-year college grades. The College Board publishes the relevant correlations. On a scale of 0 (no correlation at all) to 1 (perfect correlation), the correlation between high school grade point average and first-year college grade point average is .56. (To put that in perspective, the correlation between height and weight for adult men in the United States is about .4.) The correlation between the SAT composite score (critical reading, math, and writing) and first-year college GPA is also .56.[1] That would seem to argue for ditching the SAT, as the test does not seem to do any better at predicting college performance than high school grades. In fact, the best predictor of all is a combination of SAT scores and high school GPA, which has a correlation of .64 with first-year college grades. Sorry about that.

One crucial point in this general discussion is that correlation does not imply causation; a positive or negative association between two variables does not necessarily mean that a change in one of the variables is causing the change in the other. For example, I alluded earlier to a likely positive correlation between a student's SAT scores and the number of televisions that his family owns. This does not mean that overeager parents can boost their children's test scores by buying an extra five televisions for the house. Nor does it likely mean that watching lots of television is good for academic achievement.

The most logical explanation for such a correlation would be that highly educated parents can afford a lot of televisions and tend to have children who test better than average. Both the televisions and the test scores are likely caused by a third variable, which is parental education. I can't prove the correlation between TVs in the home and SAT scores. (The College Board does not provide such data.) However, I can prove that students in wealthy families have higher mean SAT scores than students in less wealthy families. According to the College Board, students with a family income over $200,000 have a mean SAT math score of 586, compared with a mean SAT math score of 460 for students with a family income of $20,000 or less.[2] Meanwhile, it's also likely that families with incomes over $200,000 have more televisions in their (multiple) homes than families with incomes of $20,000 or less.

I began writing this chapter many days ago. Since then, I've had a chance to watch the documentary film *Bhutto*. Wow! This is a remarkable film about a remarkable family. The original footage, stretching all the way from the partition of India and Pakistan in 1947 to the assassination of Benazir Bhutto in 2007, is extraordinary. Bhutto's voice is woven effectively throughout the film in the form of speeches and interviews. Anyway, I gave the film five stars, which is pretty much what Netflix predicted.

At the most basic level, Netflix is exploiting the concept of correlation. First, I rate a set of films. Netflix compares my ratings with those of other customers to identify those whose ratings are highly correlated with mine. Those customers tend to like the films that I like. Once that is established, Netflix can recommend films that like-minded customers have rated highly but that I have not yet seen.

That's the "big picture." The actual methodology is much more complex. In fact, Netflix launched a contest in 2006 in which members of the public were invited to design a mechanism that improved on existing Netflix recommendations by at least 10 percent (meaning that the system was 10 percent more accurate in predicting how a customer would rate a film after seeing it). The winner would get $1,000,000.

Every individual or team that registered for the contest was given "training data" consisting of more than 100 million ratings of 18,000 films by 480,000 Netflix customers. A separate set of 2.8 million ratings was "withheld," meaning that Netflix knew how the customers rated these films but the contest participants did not. The competitors were judged on how well their algorithms predicted the actual customer reviews for these withheld films. Over three years, thousands of teams from over 180 countries submitted proposals. There were two requirements for entry. First, the winner had to license the algorithm to Netflix. And second, the winner had to "describe to the world how you did it and why it works."[3]

In 2009 Netflix announced a winner: a seven-person team made up of statisticians and computer scientists from the United States, Austria, Canada, and Israel. Alas, I cannot describe the winning system, even in an appendix. The paper explaining the system is ninety-two pages long.* I'm impressed by the quality of the Netflix recommendations. Still, the system is just a super fancy variation on what people have been doing since the dawn of film: find someone with similar tastes and ask for a recommendation. You tend to like what I like, and to dislike what I dislike, so what did you think of the new George Clooney film?

That is the essence of correlation.

APPENDIX TO CHAPTER 4

To calculate the correlation coefficient between two sets of numbers, you would perform the following steps, each of which is illustrated by use of the data on heights and weights for 15 hypothetical students in the table below.

1. Convert the height of each student to standard units: (height – mean)/standard deviation.
2. Convert the weight of each student to standard units: (weight – mean)/standard deviation.
3. Calculate the product for each student of (weight in standard units) × (height in standard units). You should see that this number will be largest in absolute value when a student's height and weight are both relatively far from the mean.
4. The correlation coefficient is the sum of the products calculated above divided by the number of observations (15 in this case).

---

* You can read it at http://www.netflixprize.com/assets/GrandPrize2009_BPC_PragmaticTheory.pdf.

The correlation between height and weight for this group of students is .83. Given that the correlation coefficient can range from –1 to 1, this is a relatively high degree of positive correlation, as we would expect with height and weight.

A	B	C	D	E	F
Student	Height	Weight	Height in standard units	Weight in standard units	(Weight in standard units) × (Height in standard units)
Nick	74	193	1.21	0.99	1.19
Elana	66	133	–0.63	–0.67	0.42
Dinah	68	155	–0.17	–0.06	0.01
Rebecca	69	147	0.06	–0.29	–0.02
Ben	73	175	0.98	0.49	0.48
Charu	70	128	0.29	–0.81	–0.24
Sahar	60	100	–2.00	–1.59	3.18
Maggie	63	128	–1.32	–0.81	1.07
Faisal	67	170	–0.40	0.35	–0.14
Ted	70	182	0.29	0.68	0.20
Narciso	70	178	0.29	0.57	0.17
Katrina	70	118	0.29	–1.09	–0.32
CJ	75	227	1.44	1.93	2.77
Sophia	62	115	–1.54	–1.17	1.81
Will	74	211	1.21	1.49	1.80
Mean	68.73	157.33			Total = 12.39
Standard Deviation	4.36	36.12		Correlation coefficient = Total/n = 12.39/15 = 0.83	

The formula for calculating the correlation coefficient requires a little detour with regard to notation. The figure Σ, known as the summation sign, is a handy character in statistics. It represents the summation of the quantity that comes after it. For example, if there is a set of observations x_1, x_2, x_3, and x_4, then $\Sigma (x_i)$ tells us that we should sum the four observations: $x_1 + x_2 + x_3 + x_4$. Thus, $\Sigma (x_i) = x_1 + x_2 + x_3 + x_4$. Our formula for the mean of a set of i observations could be represented as the following: mean = $\Sigma (x_i)/n$.

We can make the formula even more adaptable by writing $\sum_{i=1}^{n} (x_i)$, which sums the quantity $x_1 + x_2 + x_3 + \ldots x_n$, or, in other words, all the terms beginning with x_1 (because i = 1) up to x_n (because i = n). Our formula for the mean of a set of n observations could be represented as the following:

$$\text{mean} = \sum_{i=1}^{n} (x_i)/_n$$

Given that general notation, the formula for calculating the correlation coefficient, r, for two variables x and y is the following:

$$r = \frac{1}{n} \sum_{i=1}^{n} \frac{(x_i - \bar{x})}{\sigma_x} \frac{(y_i - \bar{y})}{\sigma_y}$$

where

n = the number of observations;
$\bar{x}$ is the mean for variable *x*;
$\bar{y}$ is the mean for variable *y*;
σ_x is the standard deviation for variable x;
σ_y is the standard deviation for variable y.

Any statistical software program with statistical tools can also calculate the correlation coefficient between two variables. In the student height and weight example, using Microsoft Excel yields the same correlation between height and weight for the fifteen students as the hand calculation in the chart above: 0.83.

CHAPTER 5

Basic Probability

Don't buy the extended warranty on your $99 printer

In 1981, the Joseph Schlitz Brewing Company spent $1.7 million for what appeared to be a shockingly bold and risky marketing campaign for its flagging brand, Schlitz. At halftime of the Super Bowl, in front of 100 million people around the world, the company broadcast a live taste test pitting Schlitz Beer against a key competitor, Michelob.[1] Bolder yet, the company did not pick random beer drinkers to evaluate the two beers; *it picked 100 Michelob drinkers*. This was the culmination of a campaign that had run throughout the NFL playoffs.[2] There were five live television taste tests in all, each of which had 100 consumers of a competing brand (Budweiser, Miller, or Michelob) conduct a blind taste test between their supposed favorite beer and Schlitz. Each of the beer taste-offs was promoted aggressively, just like the playoff game during which it would be held (e.g., "Watch Schlitz v. Bud, Live during the AFC Playoffs").

The marketing message was clear: Even beer drinkers who think they like another brand will prefer Schlitz in a blind taste test. For the Super Bowl spot, Schlitz even hired a former NFL referee to oversee the test. Given the risky nature of conducting blind taste tests in front of huge audiences on live TV, one can assume that Schlitz produced a spectacularly delicious beer, right?

Not necessarily. Schlitz needed only a mediocre beer and a solid grasp of statistics to know that this ploy—a term I do not use lightly, even when it comes to beer advertising—would almost certainly work out in its favor. Most beers in the Schlitz category taste about the same; ironically, that is exactly the fact that this advertising campaign exploited. Assume that the typical beer drinker off the street cannot tell Schlitz from Budweiser from Michelob from Miller. In that case, a blind taste test between any two of the beers is essentially a coin flip. On average, half the taste testers will pick Schlitz, and half will pick the beer it is "challenging." This fact alone would probably *not* make a particularly effective advertising campaign. ("You can't tell the difference, so you might as well drink Schlitz.") And Schlitz absolutely, positively would not want to do this test among its own loyal customers; roughly half of these Schlitz drinkers would pick the competing beer. It looks bad when the beer drinkers supposedly most committed to your brand choose a competitor in a blind taste test—which is exactly what Schlitz was trying to do to its competitors.

Schlitz did something cleverer. The genius of the campaign was conducting the taste test exclusively among beer drinkers who stated that they preferred a competing beer. If the blind taste test is really just a coin flip, then roughly half of the Budweiser or Miller or Michelob drinkers will end up picking Schlitz. That makes Schlitz look really good. *Half of all Bud drinkers like Schlitz better!*

And it looks particularly good at halftime of the Super Bowl with a former NFL referee (in uniform) conducting the taste test. Still, it's live television. Even if the statisticians at Schlitz had determined with loads of previous private trials that the typical Michelob drinker will pick Schlitz 50 percent of the time, what if the 100 Michelob drinkers taking the test at halftime of the Super Bowl turn out to be quirky? Yes, the blind taste test is the equivalent of a coin toss, but what if most of the tasters chose Michelob *just by chance*? After all, if we lined up the same 100 guys and asked them to flip a coin, it's entirely possible that they would flip 85 or 90 tails. That kind of bad luck in the taste test would be a disaster for the Schlitz brand (not to mention a waste of the $1.7 million for the live television coverage).

Statistics to the rescue! If there were some kind of statistics superhero,* this is when he or she would have swooped into the Schlitz corporate headquarters and unveiled the details of what statisticians call a binomial experiment (also called a Bernoulli trial). The key characteristics of a binomial experiment are that we have a fixed number of trials (e.g., 100 taste testers), each with two possible outcomes (Schlitz or Michelob), and the probability of "success" is the same in each trial. (I am assuming the probability of picking one beer or the other is 50 percent, and I am defining "success" as a tester picking Schlitz.) We also assume that all the "trials" are independent, meaning that one blind taste tester's decision has no impact on any other tester's decision.

With only this information, a statistical superhero can calculate the probability of all the different outcomes for the 100 trials, such as 52 Schlitz and 48 Michelob or 31 Schlitz and 69 Michelob. Those of us who are not statistical superheroes can use a computer to do the same thing. The chances of all 100 taste testers picking Michelob were 1 in 1,267,650,600,228,229,401,496,703,205,376. There was probably a bigger chance that all of the testers would be killed at halftime by an asteroid. More important, the same basic calculations can give us the cumulative probability for a range of outcomes, such as the chances that 40 or fewer testers pick Schlitz. These numbers would clearly have assuaged the fears of the Schlitz marketing folks.

Let's assume that Schlitz would have been pleased if at least 40 of the 100 tasters picked Schlitz—an impressive number given that all of the men taking the live blind taste test had professed to be Michelob drinkers. An outcome *at least that good* was highly likely. If the taste test is really like a flip of the coin, then basic probability tells us that there was a 98 percent chance that at least 40 of the tasters would pick Schlitz, and an 86 percent chance that at least 45 of the tasters would.† In theory, this wasn't a very risky gambit at all.

---

* I have in mind "Six Sigma Man." The lowercase Greek letter sigma, σ, represents the standard deviation. Six Sigma Man is six standard deviations above the norm in terms of statistical ability, strength, and intelligence.

† For all of these calculations, I've used a handy online binomial calculator, at http://stattrek.com/Tables/Binomial.aspx.

So what happened to Schlitz? At halftime of the 1981 Super Bowl, exactly 50 percent of the Michelob drinkers chose Schlitz in the blind taste test.

There are two important lessons here: probability is a remarkably powerful tool, and many leading beers in the 1980s were indistinguishable from one another. This chapter will focus primarily on the first lesson.

Probability is the study of events and outcomes involving an element of uncertainty. Investing in the stock market involves uncertainty. So does flipping a coin, which may come up heads or tails. Flipping a coin four times in a row involves additional layers of uncertainty, because each of the four flips can result in a head or a tail. If you flip a coin four times in a row, I cannot know the outcome in advance with certainty (nor can you). Yet I *can* determine in advance that some outcomes (two heads, two tails) are more likely than others (four heads). As the folks at Schlitz reckoned, those kinds of probability-based insights can be extremely helpful. In fact, if you can understand why the probability of flipping four heads in a row with a fair coin is 1 in 16, you can (with some work) understand everything from how the insurance industry works to whether a pro football team should kick the extra point after a touchdown or go for a two-point conversion.

Let's start with the easy part: Many events have known probabilities. The probability of flipping heads with a fair coin is ½. The probability of rolling a one with a single die is ⅙. Other events have probabilities that can be inferred on the basis of past data. The probability of successfully kicking the extra point after touchdown in professional football is .94, meaning that kickers make, on average, 94 out of every 100 extra-point attempts. (Obviously this figure might vary slightly for different kickers, under different weather circumstances, and so on, but it's not going to change radically.) Simply having and appreciating this kind of information can often clarify decision making and render risks explicit. For example, the Australian Transport Safety Board published a report quantifying the fatality risks for different modes of transport. Despite widespread fear of flying, the risks associated with commercial air travel are tiny. Australia hasn't had a commercial air fatality since the 1960s, so the fatality rate per 100 million kilometers traveled is essentially zero.

The rate for drivers is .5 fatalities per 100 million kilometers traveled. The really impressive number is for motorcycles—if you aspire to be an organ donor. The fatality rate is thirty-five times higher for motorcycles than for cars.[3]

In September of 2011, a 6.5-ton NASA satellite was plummeting to earth and was expected to break apart once it hit the earth's atmosphere. What were the chances of being struck by the debris? Should I have kept the kids home from school? The rocket scientists at NASA estimated that the probability of any individual person's being hit by a part of the falling satellite was 1 in 21 trillion. Yet the chances that anyone anywhere on earth might get hit were a more sobering 1 in 3,200.* In the end, the satellite did break apart on reentry, but scientists aren't entirely certain where all the pieces ended up.[4] No one reported being hurt. Probabilities do not tell us what will happen for sure; they tell us what is *likely to happen* and what is *less likely to happen*. Sensible people can make use of these kinds of numbers in business and life. For example, when you hear on the radio that a satellite is plummeting to earth, you should not race home on your motorcycle to warn the family.

When it comes to risk, our fears do not always track with what the numbers tell us we should be afraid of. One of the striking findings from *Freakonomics*, by Steve Levitt and Stephen Dubner, was that swimming pools in the backyard are far more dangerous than guns in the closet.[5] Levitt and Dubner calculate that a child under ten is one hundred times more likely to die in a swimming pool than from a gun accident.† An intriguing paper by three Cornell researchers, Garrick Blalock, Vrinda

* NASA also pointed out that even falling space debris is government property. Apparently it is illegal to keep a satellite souvenir, even if it lands in your backyard.

† The Levitt and Dubner calculations are as follows. Each year roughly 550 children under ten drown and 175 children under ten die from gun accidents. The rates they compare are 1 drowning for every 11,000 residential pools compared with 1 gun death per "million-plus" guns. For adolescents, I suspect the numbers may change sharply, both because they are better able to swim and because they are more likely to cause a tragedy if they stumble upon a loaded gun. However, I have not checked the data on this point.

Kadiyali, and Daniel Simon, found that thousands of Americans may have died since the September 11 attacks *because they were afraid to fly*.[6] We will never know the true risks associated with terrorism; we do know that driving is dangerous. When more Americans opted to drive rather than to fly after 9/11, there were an estimated 344 additional traffic deaths per month in October, November, and December of 2001 (taking into account the average number of fatalities and other factors that typically contribute to road accidents, such as weather). This effect dissipated over time, presumably as the fear of terrorism diminished, but the authors of the study estimate that the September 11 attacks may have caused more than 2,000 driving deaths.

Probability can also sometimes tell us *after the fact* what likely happened and what likely did not happen—as in the case of DNA analysis. When the technicians on *CSI: Miami* find a trace of saliva on an apple core near a murder victim, that saliva does not have the murderer's name on it, even when viewed under a powerful microscope by a very attractive technician. Instead, the saliva (or hair, or skin, or bone fragment) will contain a DNA segment. Each DNA segment in turn has regions, or loci, that can vary from individual to individual (except for identical twins, who share the same DNA). When the medical examiner reports that a DNA sample is a "match," that's only part of what the prosecution has to prove. Yes, the loci tested on the DNA sample from the crime scene must match the loci on the DNA sample taken from the suspect. However, the prosecutors must also prove that the match between the two DNA samples is not merely a coincidence.

Humans share similarities in their DNA, just as we share other similarities: shoe size, height, eye color. (More than 99 percent of all DNA is identical among all humans.) If researchers have access to only a small sample of DNA on which only a few loci can be tested, it's possible that thousands or even millions of individuals may share that genetic fragment. Therefore, the more loci that can be tested, and the more natural genetic variation there is in each of those loci, the more certain the match becomes. Or, to put it a bit differently, the less likely it becomes that the DNA sample will match more than one person.[7]

To get your mind around this, imagine that your "DNA number" consists of your phone number attached to your Social Security number. This nineteen-digit sequence uniquely identifies you. Consider each digit a "locus" with ten possibilities: 0, 1, 2, 3, and so on. Now suppose that crime scene investigators find the remnant of a "DNA number" at a crime scene: _ _ 4 5 9 _ _ _ 4 _ 0 _ 9 8 1 7 _ _ _. This happens to match exactly with your "DNA number." Are you guilty?

You should see three things. First, anything less than a full match of the entire genome leaves some room for uncertainty. Second, the more "loci" that can be tested, the less uncertainty remains. And third, context matters. This match would be extremely compelling if you also happened to be caught speeding away from the crime scene with the victim's credit cards in your pocket.

When researchers have unlimited time and resources, the typical process involves testing thirteen different loci. The chances that two people share the same DNA profile across all thirteen loci are extremely low. When DNA was used to identify the remains found in the World Trade Center after September 11, samples found at the scene were matched to samples provided by family members of the victims. The probability required to establish positive identification was one in a billion, meaning that the probability that the discovered remains belonged to someone other than the identified victim had to be judged as one in one billion or less. Later in the search, this standard was relaxed, as there were fewer unidentified victims with whom the remains could be confused.

When resources are limited, or the available DNA sample is too small or too contaminated for thirteen loci to be tested, things get more interesting and controversial. The *Los Angeles Times* ran a series in 2008 examining the use of DNA as criminal evidence.[8] In particular, the *Times* questioned whether the probabilities typically used by law enforcement understate the likelihood of coincidental matches. (Since no one knows the DNA profile of the entire population, the probabilities presented in court by the FBI and other law enforcement entities are estimates.) The intellectual pushback was instigated when a crime lab analyst in Arizona running tests with the state's DNA database discovered two unrelated felons whose DNA matched at nine loci; according to the FBI, the chances of a nine-loci match between two unrelated persons are 1 in 113 billion. Subsequent searches of other DNA databases turned up more than a thousand human pairs with genetic matches at nine loci or more. I'll leave this issue for law enforcement and defense lawyers to work out. For now, the lesson is that the dazzling science of DNA analysis is only as good as the probabilities used to support it.

Often it is extremely valuable to know the likelihood of multiple events' happening. What is the probability that the electricity goes out *and* the generator doesn't work? The probability of two independent events' *both* happening is the product of their respective probabilities. In other words, the probability of Event A happening *and* Event B happening is the probability of Event A multiplied by the probability of Event B. An example makes it much more intuitive. If the probability of flipping heads with a fair coin is ½, then the probability of flipping heads twice in a row is ½ × ½, or ¼. The probability of flipping three heads in a row is ⅛, the probability of four heads in a row is 1/16, and so on. (You should see that the probability of throwing four tails in a row is also 1/16.) This explains why the system administrator at your school or office is constantly on your case to improve the "quality" of your password. If you have a six-digit password using only numerical digits, we can calculate the number of possible passwords: 10 × 10 × 10 × 10 × 10 × 10, which equals 10^6, or 1,000,000. That sounds like a lot of possibilities, but a computer could blow through all 1,000,000 possible combinations in a fraction of a second.

So let's suppose that your system administrator harangues you long enough that you include letters in your password. At that point, each of the 6 digits now has 36 combinations: 26 letters and 10 digits. The number of possible passwords grows to 36 × 36 × 36 × 36 × 36 × 36, or 36^6, which is over two billion. If your administrator demands eight digits and urges you to use symbols like #, @, % and !, as the University of Chicago does, the number of potential passwords climbs to 46^8, or just over 20 trillion.

There is one crucial distinction here. This formula is applicable only if the events are independent, meaning that the outcome of one has no effect on the outcome of another. For example, the probability that you throw heads on the first flip does not change the likelihood of your throwing heads on the second flip. On the other hand, the probability that it rains today is *not* independent of whether it rained yesterday, since storm fronts can last for days. Similarly, the probability of crashing your car today and crashing your car next year are not independent. Whatever caused you to crash this year might also cause you to crash next year; you might be prone to drunk driving, drag racing, texting while driving, or just driving badly. (This is why your auto insurance rates go up after an accident; it is not simply that the company wants to recover the money that it has paid out for the claim; rather, it now has new information about your probability of crashing in the future, which—after you've driven the car through your garage door—has gone up.)

Suppose you are interested in the probability that one event happens *or* another event happens: outcome A *or* outcome B (again assuming that they are independent). In this case, the probability of getting A or B consists of the sum of their individual probabilities: the probability of A *plus* the probability of B. For example, the likelihood of throwing a 1, 2, or 3, with a single die is the sum of their individual probabilities: $\frac{1}{6} + \frac{1}{6} + \frac{1}{6} = \frac{3}{6} = \frac{1}{2}$. This should make intuitive sense. There are six possible outcomes for the roll of a die. The numbers 1, 2, and 3 collectively make up half of those possible outcomes. Therefore you have a 50 percent chance of rolling a 1, 2, or 3. If you are playing craps in Las Vegas, the chance of rolling a 7 or 11 in a single throw is the number of combinations that sum to 7 or 11 divided by the total number of combinations that can be thrown with two dice, or $\frac{8}{36}$.*

Probability also enables us to calculate what might be the most

* There are 6 ways to throw a 7 with two dice: (1,6); (2,5); (3,4); (6,1); (5,2); and (4,3). There are only 2 ways to throw an 11: (5,6) and (6,5).

Meanwhile, there are 36 total possible throws with two dice: (1,1); (1,2); (1,3); (1,4); (1,5); (1,6). And (2,1); (2,2); (2,3); (2,4); (2,5); (2,6). And (3,1); (3,2); (3,3); (3,4); (3,5); (3,6). And (4,1); (4,2); (4,3); (4,4); (4,5); (4,6). And (5,1); (5,2); (5,3); (5,4); (5,5); (5,6). And, finally, (6,1); (6,2); (6,3); (6,4); (6,5); and (6,6).

Thus, the chance of throwing a 7 or 11 is the number of possible ways of throwing either of those two numbers divided by the total number of possible throws with two dice, which is 8/36. Incidentally, much of the earlier research on probability was done by gamblers to determine exactly this kind of thing.

useful tool in all of managerial decision making, particularly finance: expected value. The expected value takes basic probability one step further. The expected value or payoff from some event, say purchasing a lottery ticket, is the sum of all the different outcomes, each weighted by its probability and payoff. As usual, an example makes this clearer. Suppose you are invited to play a game in which you roll a single die. The payoff to this game is $1 if you roll a 1; $2 if you roll a 2; $3 if you roll a 3; and so on. What is the expected value for a single roll of the die? Each possible outcome has a 1/6 probability, so the expected value is:

$\frac{1}{6}(\$1) + \frac{1}{6}(\$2) + \frac{1}{6}(\$3) + \frac{1}{6}(\$4) + \frac{1}{6}(\$5) + \frac{1}{6}(\$6) = \frac{21}{6}$, or $3.50.

At first glance, the expected value of $3.50 might appear to be a relatively useless figure. After all, you can't actually earn $3.50 with a single roll of the die (since your payoff has to be a whole number). In fact, the expected value turns out to be extremely powerful because it can tell you whether a particular event is "fair," given its price and expected outcome. Suppose you have the chance to play the above game for $3 a throw. Does it make sense to play? Yes, because the expected value of the outcome ($3.50) is higher than the cost of playing ($3.00). This does not guarantee that you will make money by playing once, but it does help clarify which risks are worth taking and which are not.

We can take this hypothetical example and apply it to professional football. As noted earlier, after a touchdown, teams have a choice between kicking an extra point and attempting a two-point conversion. The former involves kicking the ball through the goalposts from the three yard line; the latter involves running or passing it into the end zone from the three yard line, which is significantly more difficult. Teams can choose the easy option and get one point, or they can choose the harder option and get two points. What to do?

Statisticians may not play football or date cheerleaders, but they can provide statistical guidance for football coaches.[9] As pointed out earlier, the probability of making the kick after a touchdown is .94. This means that the expected value of a point-after attempt is also .94, since it equals the payoff (1 point) multiplied by the probability of success (.94). No team ever scores .94 points, but this figure is helpful in quantifying the value of attempting this option after a touchdown relative to the alternative, which is the two-point conversion. The expected value of "going for two" is much lower: .74. Yes, the payoff is higher (2 points), but the success rate is dramatically lower (.37). Obviously if there is one second left in the game and a team is behind by two points after scoring a touchdown, it has no choice but to go for a two-point conversion. But if a team's goal is to maximize points scored over time, then kicking the extra point is the strategy that will do that.

The same basic analysis can illustrate why you should never buy a lottery ticket. In Illinois, the probabilities associated with the various possible payoffs for the game are printed on the back of each ticket. I purchased a $1 instant ticket. (Note to self: Is this tax deductible?) On the back—in tiny, tiny print—are the chances of winning different cash prizes, or a free new ticket: 1 in 10 (free ticket); 1 in 15 ($2); 1 in 42.86 ($4); 1 in 75 ($5); and so on up to the 1 in 40,000 chance of winning $1,000. I calculated the expected payout for my instant ticket by adding up each possible cash prize weighted by its probability.* It turns out that my $1 lottery ticket has an expected payout of roughly $.56, making it an absolutely miserable way to spend $1. As luck would have it, I won $2.

My $2 prize notwithstanding, buying the ticket was a stupid thing to do. This is one of the crucial lessons of probability. Good decisions—as measured by the underlying probabilities—can turn out badly. And bad decisions—like spending $1 on the Illinois lottery—can still turn out well, at least in the short run. But probability triumphs in the end. An important theorem known as the law of large numbers tells us that as the number of trials increases, the average of the outcomes will get closer and closer to its expected value. Yes, I won $2 playing the lotto

---

* The full expected value for the Illinois Dugout Doubler $1 ticket (rounded to the nearest cent) is as follows: 1/15 ($2) + 1/42.86 ($4) + 1/75 ($5) + 1/200 ($10) + 1/300 ($25) + 1/1,589.40 ($50) + 1/8000 ($100) + 1/16,000 ($200) + 1/48,000 ($500) + 1/40,000 ($1,000) = $.13 + $.09 + $.07 + $.05 + $.08 + $.03 + $.01 + $.01 + $.01 + $.03 = $.51. However, there is also a 1/10 chance of getting a free ticket, which has an expected payout of $.51, so the overall expected payout is $.51 + .1 ($.51) = $.51 + $.05 = $.56.

today. And I might win $2 again tomorrow. But if I buy thousands of $1 lottery tickets, each with an expected payout of $.56, then it becomes a near mathematical certainty that I will lose money. By the time I've spent $1 million on tickets, I'm going to end up with something strikingly close to $560,000.

The law of large numbers explains why casinos always make money in the long run. The probabilities associated with all casino games favor the house (assuming that the casino can successfully prevent blackjack players from counting cards). If enough bets are wagered over a long enough time, the casino will be certain to win more than it loses. The law of large numbers also demonstrates why Schlitz was much better off doing 100 blind taste tests at halftime of the Super Bowl rather than just 10. Check out the "probability density functions" for a Schlitz type of test with 10, 100, and 1,000 trials. (Although it sounds fancy, a probability density function merely plots the assorted outcomes along the *x*-axis and the expected probability of each outcome on the *y*-axis; the weighted probabilities—each outcome multiplied by its expected frequency—will add up to 1.) Again I'm assuming that the taste test is just like a coin flip and each tester has a .5 probability of choosing Schlitz. As you can see below, the expected outcome converges around 50 percent of tasters' choosing Schlitz as the number of tasters gets larger. At the same time, the probability of getting an outcome that deviates sharply from 50 percent falls sharply as the number of trials gets large.

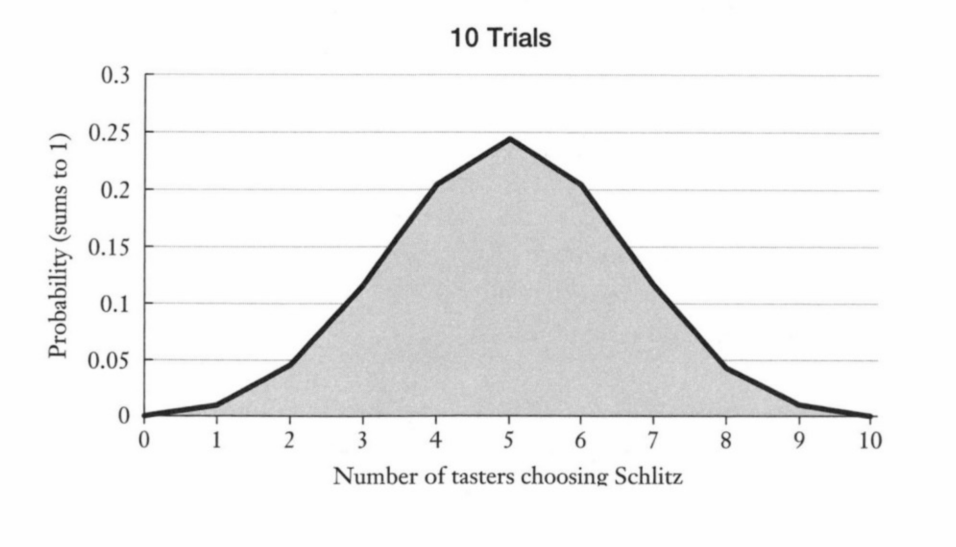

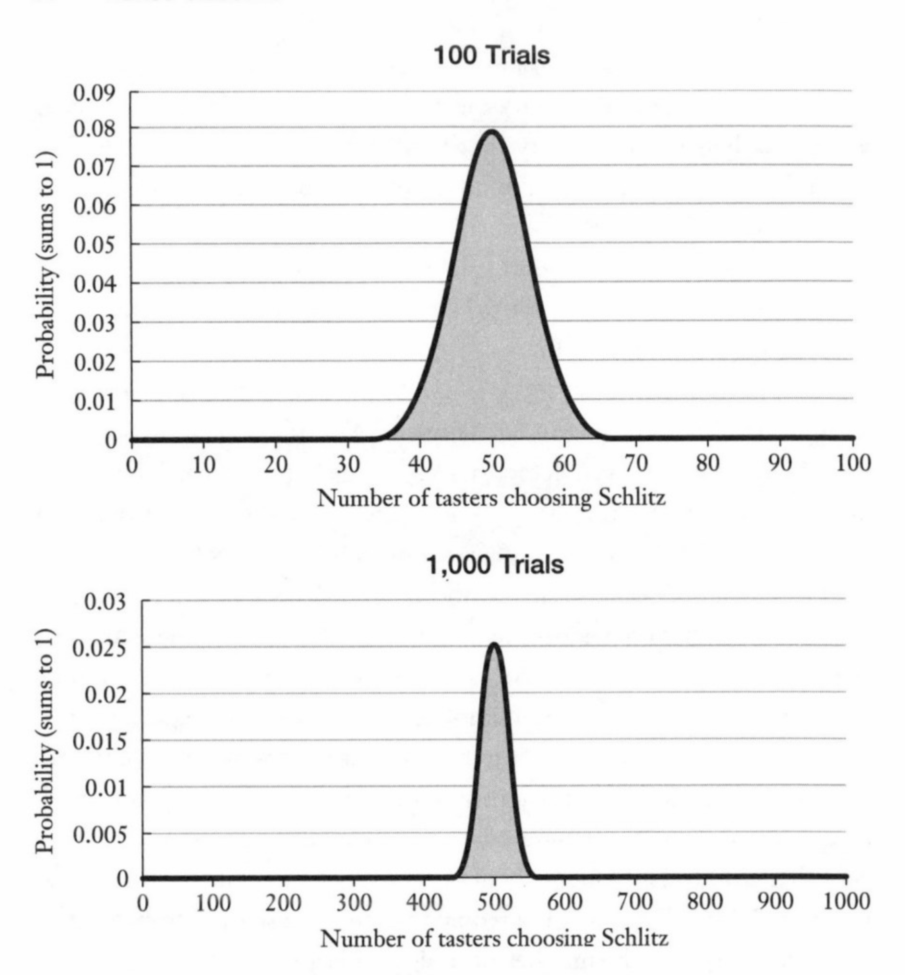

I stipulated earlier that Schlitz executives would be happy if 40 percent or more of the Michelob drinkers chose Schlitz in the blind taste test. The figures below reflect the probability of getting that outcome as the number of tasters gets larger:

10 blind taste testers: .83
100 blind taste testers: .98
1,000 blind taste testers: .9999999999
1,000,000 blind taste testers: 1

By now the intuition is obvious behind the chapter subtitle, "Don't buy the extended warranty on your $99 printer." Okay, maybe that's not so obvious. Let me back up. The entire insurance industry is built on probability. (A warranty is just a form of insurance.) When you insure anything, you are contracting to receive some specified payoff in the event of a clearly defined contingency. For example, your auto insurance will replace your car in the event that it gets stolen or crushed by a tree. In exchange for this guarantee, you agree to pay some fixed amount of money for the period in which you are insured. The key idea is that in exchange for a regular and predictable payment, you have transferred to the insurance company the risk of having your car stolen, crushed, or even totaled by your own bad driving.

Why are these companies willing to assume such risks? Because they will earn large profits in the long run if they price their premiums correctly. Obviously some cars insured by Allstate will get stolen. Others will get totaled when their owners drive over a fire hydrant, as happened to my high school girlfriend. (She also had to replace the fire hydrant, which is far more expensive than you might think.) But most cars insured by Allstate or any other company will be just fine. To make money, the insurance company need only collect more in premiums than it pays out in claims. And to do that, the firm must have a solid grasp of what is known in industry jargon as the "expected loss" on every policy. This is exactly the same concept as expected value, only with an insurance twist. If your car is insured for $40,000, and the chances of its getting stolen in any given year are 1 in 1,000, then the annual expected loss on your car is $40. The annual premium for the theft portion of the coverage needs to be *more than $40.* At that point, the insurance company becomes just like the casino or the Illinois lottery. Yes, there will be payouts, but over the long run what comes in will be more than what goes out.

As a consumer, you should recognize that insurance *will not* save you money in the long run. What it *will* do is prevent some unacceptably high loss, such as replacing a $40,000 car that was stolen or a $350,000 house that burned down. Buying insurance is a "bad bet" from a statistical standpoint since you will pay the insurance company, on average, more than you get back. Yet it can still be a sensible tool for protecting against outcomes that would otherwise wreck your life. Ironically, someone as rich as Warren Buffett can save money by not purchasing car insurance, homeowner's insurance, or even health insurance because he can afford whatever bad things might happen to him.

Which finally brings us back to your $99 printer! We'll assume that you've just picked out the perfect new laser printer at Best Buy or some other retailer.* When you reach the checkout counter, the sales assistant will offer you a series of extended warranty options. For another $25 or $50, Best Buy will fix or replace the printer should it break in the next year or two. On the basis of your understanding of probability, insurance, and basic economics, you should immediately be able to surmise all of the following: (1) Best Buy is a for-profit business that seeks to maximize profits. (2) The sales assistant is eager for you to buy the extended warranty. (3) From numbers 1 and 2, we can infer that the cost of the warranty to you is greater than the expected cost of fixing or repairing the printer for Best Buy. If this were not the case, Best Buy would not be so aggressive in trying to sell it to you. (4) If your $99 printer breaks and you have to pay out of pocket to fix or replace it, this will not meaningfully change your life.

On average, you'll pay more for the extended warranty than you would to repair the printer. The broader lesson—and one of the core lessons of personal finance—is that you should always insure yourself against any adverse contingency that you cannot comfortably afford to withstand. You should skip buying insurance on everything else.

Expected value can also help us untangle complex decisions that involve many contingencies at different points in time. Suppose a friend of yours has asked you to invest $1 million in a research venture examining a new cure for male pattern baldness. You would probably ask what the likelihood of success will be; you'll get a complicated answer. This is a research project, so there is only a 30 percent chance that the team will discover a cure that works. If the team does not find a cure, you will get $250,000 of your investment back, as those funds will have been reserved for taking the drug to market (testing, marketing, etc.) Even if the researchers are

* Earlier in the book I used an example that involved drunken employees producing defective laser printers. You will need to forget that example here and assume that the company has fixed its quality problems.

successful, there is only a 60 percent chance that the U.S. Food and Drug Administration will approve the new miracle baldness cure as safe for use on humans. Even then, if the drug is safe and effective, there is a 10 percent chance that a competitor will come to market with a better drug at about the same time, wiping out any potential profits. If everything goes well—the drug is safe, effective, and unchallenged by competitors—then the best estimate on the return on your investment is $25 million.

Should you make the investment?

This seems like a muddle of information. The potential payday is huge—25 times your initial investment—but there are so many potential pitfalls. A decision tree can help organize this kind of information and—if the probabilities associated with each outcome are correct—give you a probabilistic assessment of what you ought to do. The decision tree maps out each source of uncertainty and the probabilities associated with all possible outcomes. The end of the tree gives us all the possible payoffs and the probability of each. If we weight each payoff by its likelihood, and sum all the possibilities, we will get the expected value of this investment opportunity. As usual, the best way to understand this is to take a look.

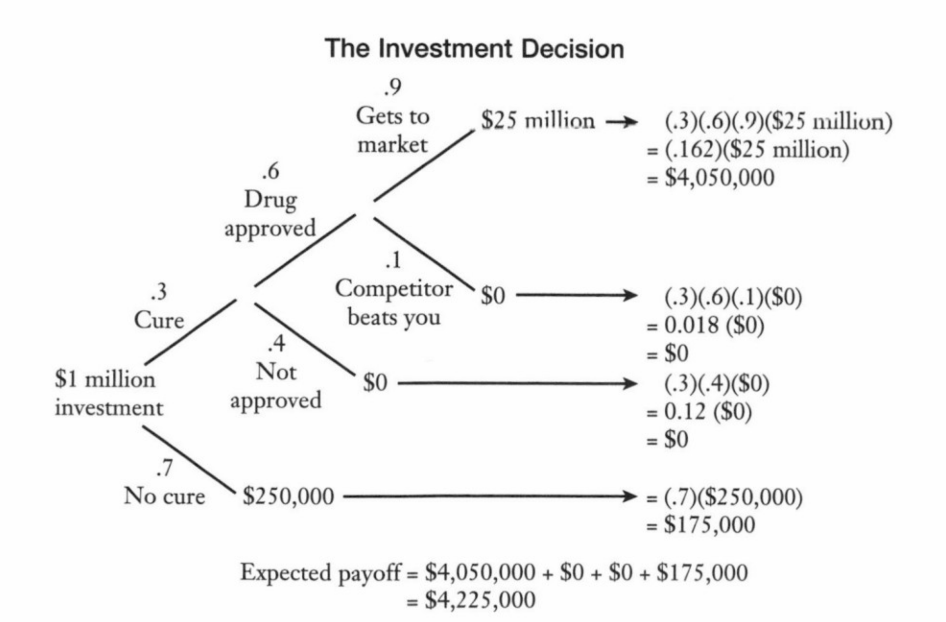

This particular opportunity has an attractive expected value. The weighted payoff is $4.225 million. Still, this investment may not be the wisest thing to do with the college tuition money that you've squirreled away for your children. The decision tree lets you know that your expected payoff is far higher than what you are being asked to invest. On the other hand, the most likely outcome, meaning the one that will happen most often, is that the company will not discover a cure for baldness and you will get only $250,000 back. Your appetite for this investment might depend on your risk profile. The law of large numbers suggests that an investment firm, or a rich individual like Warren Buffet, should seek out hundreds of opportunities like this with uncertain outcomes but attractive expected returns. Some will work; many won't. On average, these investors will make a lot of money, just like an insurance company or a casino. If the expected payoff is in your favor, more trials are always better.

The same basic process can be used to explain a seemingly counterintuitive phenomenon. Sometimes it does not make sense to screen the entire population for a rare but serious disease, such as HIV/AIDS. Suppose we can test for some rare disease with a high degree of accuracy. For the sake of example, let's assume the disease affects 1 of every 100,000 adults and the test is 99.9999 percent accurate. The test never generates a false negative (meaning that it never misses someone who has the disease); however, roughly 1 in 10,000 tests conducted on a healthy person will generate a false positive, meaning that the person tests positive but does not actually have the disease. The striking outcome here is that despite the impressive accuracy of the test, *most of the people who test positive will not have the disease*. This will generate enormous anxiety among those who falsely test positive; it can also waste finite health care resources on follow-up tests and treatment.

If we test the entire American adult population, or roughly 175 million people, the decision tree looks like the following:

Widespread Screening for a Rare Disease

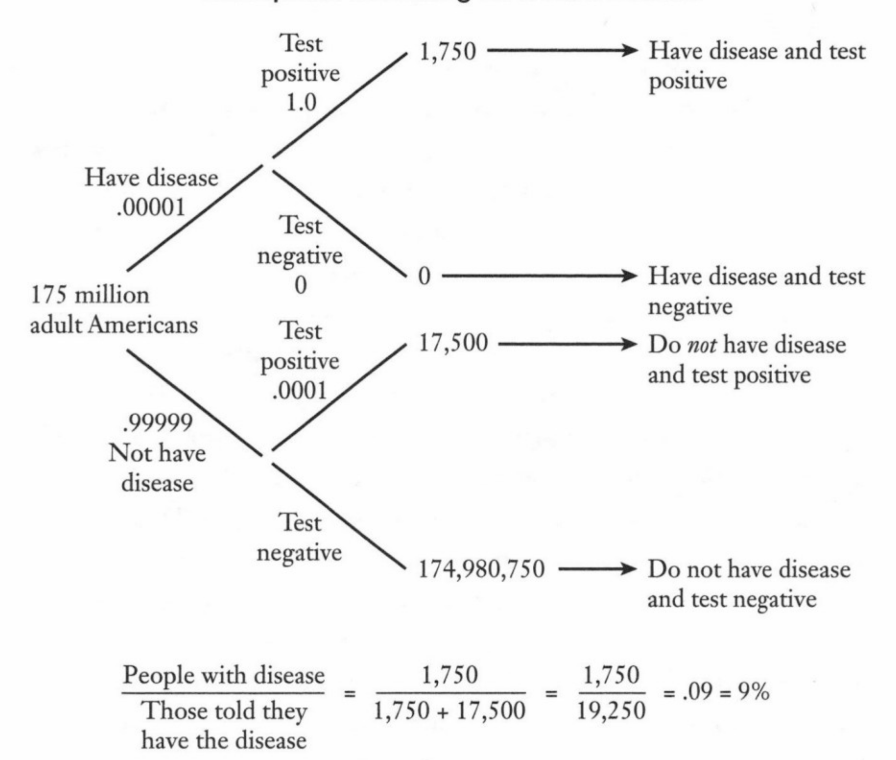

Only 1,750 adults have the disease. They all test positive. Over 174 million adults do not have the disease. Of this healthy group who are tested, 99.9999 get the correct result that they do not have the disease. Only .0001 get a false positive. But .0001 of 174 million is still a big number. In fact, 17,500 people will, on average, get false positives.

Let's look at what that means. A total of 19,250 people are notified that they have the disease; only 9 percent of them are actually sick! And that's with a test that has a very low rate of false positives. Without going too far off topic, this should give you some insight into why cost containment in health care sometimes involves less screening of healthy people for diseases, not more. In the case of a disease like HIV/AIDS, public health officials will often recommend that the resources available be used to screen the populations at highest risk, such as gay men or intravenous drug users.

• • •

Sometimes probability helps us by flagging suspicious patterns. Chapter 1 introduced the problem of institutionalized cheating on standardized tests and one of the firms that roots it out, Caveon Test Security. The Securities and Exchange Commission (SEC), the government agency responsible for enforcing federal laws related to securities trading, uses a similar methodology for catching inside traders. (Inside trading involves illegally using private information, such as a law firm's knowledge of an impending corporate acquisition, to trade stock or other securities in the affected companies.) The SEC uses powerful computers to scrutinize hundreds of millions of stock trades and look for suspicious activity, such as a big purchase of shares in a company just before a takeover is announced, or the dumping of shares just before a company announces disappointing earnings.[10] The SEC will also investigate investment managers with unusually high returns over long periods of time. (Both economic theory and historical data suggest that it is extremely difficult for a single investor to get above-average returns year after year.) Of course, smart investors are always trying to anticipate good and bad news and to devise perfectly legal strategies that consistently beat the market. Being a good investor does not necessarily make one a criminal. How does a computer tell the difference? I called the enforcement division of the SEC several times to ask what particular patterns are most likely to signal criminal activity. They still have not called me back.

In the 2002 film *Minority Report*, Tom Cruise plays a "pre-crime" detective who is part of a bureau that uses technology to predict crimes *before* they're committed.

Well folks, that's not science fiction anymore. In 2011, the *New York Times* ran the following headline: "Sending the Police before There's a Crime."[11] The story described how detectives were dispatched to a parking garage in downtown Santa Cruz by a computer program that predicted that there was a high likelihood of burglaries from cars at that location on that day. Police subsequently arrested two women peering into car windows. One had outstanding arrest warrants; the other was carrying illegal drugs.

The Santa Cruz system was designed by two mathematicians, an anthropologist, and a criminologist. The Chicago Police Department has created an entire predictive analytics unit, in part because gang activity, the source of much of the city's violence, follows certain patterns. The book *Data Mining and Predictive Analysis: Intelligence Gathering and Crime Analysis*, a guide to statistics for law enforcement, begins enthusiastically, "It is now possible to predict the future when it comes to crime, such as identifying crime trends, anticipating hotspots in the community, refining resource deployment decisions, and ensuring the greatest protection for citizens in the most efficient manner." (Look, I read these kinds of things so that you don't have to.)

"Predictive policing" is part of a broader movement called predictive analytics. Crime will always involve an element of uncertainty, as will determining who is going to crash his car or default on her mortgage. Probability helps us navigate those risks. And information refines our understanding of the relevant probabilities. Businesses facing uncertainty have always sought to quantify their risks. Lenders request things like income verification and a credit score. Yet these blunt credit instruments are starting to feel like the prediction equivalent of a caveman's stone tools. The confluence of huge amounts of digital data and cheap computing power has generated fascinating insights into human behavior. Insurance officials correctly describe their business as the "transfer of risk"—and so they had better understand the risks being transferred to them. Companies like Allstate are in the business of knowing things that might otherwise seem like random trivia:[12]

- Twenty to twenty-four-year-old drivers are the most likely to be involved in a fatal crash.
- The most commonly stolen car in Illinois is the Honda Civic (as opposed to full-size Chevrolet pickups in Alabama).*

---

* Since I've admonished you to be a stickler about descriptive statistics, I feel compelled to point out that the most commonly stolen car is not necessarily the kind of car that is most likely to be stolen. A high number of Honda Civics are reported stolen because there are a lot of them on the road; the chances that any individual Honda

- Texting while driving causes crashes, but state laws banning the practice do not seem to stop drivers from doing it. In fact, such laws might even make things worse by prompting drivers to hide their phones and therefore take their eyes off the road while texting.

The credit card companies are at the forefront of this kind of analysis, both because they are privy to so much data on our spending habits and because their business model depends so heavily on finding customers who are just barely a good credit risk. (The customers who are the best credit risks tend to be money losers because they pay their bills in full each month; the customers who carry large balances at high interest rates are the ones who generate fat profits—as long as they don't default.) One of the most intriguing studies of who is likely to pay a bill and who is likely to walk away was generated by J. P. Martin, "a math-loving executive" at Canadian Tire, a large retailer that sells a wide range of automotive products and other retail goods.[13] When Martin analyzed the data—every transaction using a Canadian Tire credit card from the prior year—he discovered that what customers purchased was a remarkably precise predictor of their subsequent payment behavior when used in conjunction with traditional tools like income and credit history.

A *New York Times Magazine* article entitled "What Does Your Credit Card Company Know about You?" described some of Martin's most intriguing findings: "People who bought cheap, generic automotive oil were much more likely to miss a credit-card payment than someone who got the expensive, name-brand stuff. People who bought carbon-monoxide monitors for their homes or those little felt pads that stop chair legs from scratching the floor almost never missed payments. Anyone who purchased a chrome-skull car accessory or a 'Mega Thruster Exhaust System' was pretty likely to miss paying his bill eventually."

Civic is stolen (which is what car insurance companies care about) might be quite low. In contrast, even if 99 percent of all Ferraris are stolen, Ferrari would not make the "most commonly stolen" list, because there are not that many of them to steal.

• • •

Probability gives us tools for dealing with life's uncertainties. You shouldn't play the lottery. You should invest in the stock market if you have a long investment horizon (because stocks typically have the best long-term returns). You should buy insurance for some things, but not others. Probability can even help you maximize your winnings on game shows (as the next chapter will show.)

That said (or written), probability is not deterministic. No, you shouldn't buy a lottery ticket—but you still might win money if you do. And yes, probability can help us catch cheaters and criminals—but when used inappropriately it can also send innocent people to jail. That's why we have Chapter 6.

CHAPTER 5½

The Monty Hall Problem

The "Monty Hall problem" is a famous probability-related conundrum faced by participants on the game show *Let's Make a Deal*, which premiered in the United States in 1963 and is still running in some markets around the world. (I remember watching the show whenever I was home sick from elementary school.) The program's gift to statisticians was described in the introduction. At the end of each day's show a contestant was invited to stand with host Monty Hall facing three big doors: Door no. 1, Door no. 2, and Door no. 3. Monty explained to the contestant that there was a highly desirable prize behind one of the doors and a goat behind the other two doors. The player chose one of the three doors and would get as a prize whatever was behind it. (I don't know if the participants actually got to keep the goat; for our purposes, assume that most players preferred the new car.)

The initial probability of winning was straightforward. There were two goats and one car. As the participant stood facing the doors with Monty, he or she had a 1 in 3 chance of choosing the door that would be opened to reveal the car. But as noted earlier, *Let's Make a Deal* had a twist, which is why the show and its host have been immortalized in the probability literature. After the contestant chose a door, Monty would open one of the two doors that the contestant *had not picked*, always revealing a goat. At that point, Monty would ask the contestant if he would like to change his pick—to switch from the closed door that he had picked originally to the other remaining closed door.

For the sake of example, assume that the contestant has originally chosen Door no. 1. Monty would then open Door no. 3; a live goat would be standing there on a stage. Two doors would still be closed, nos. 1 and 2. If the valuable prize was behind no. 1, the contestant would win; if it was behind no. 2, he would lose. That's when Monty would turn to the player and ask whether he would like to change his mind and switch doors, from no. 1 to no. 2 in this case. Remember, both doors are still closed. The only new information the contestant has received is that a goat showed up behind one of the doors that he did not pick.

Should he switch?

Yes. The contestant has a 1/3 chance of winning if he sticks with his initial choice and a 2/3 chance of winning if he switches. If you don't believe me, read on.

I'll concede that this answer seems entirely unintuitive at first. It would appear that the contestant has a one-third chance of winning no matter what he does. There are three closed doors. At the beginning, each door has a one in three chance of holding the valuable prize. How could it matter whether he switches from one closed door to another?

The answer lies in the fact that Monty Hall knows what is behind each door. If the contestant picks Door no. 1 and there is a car behind it, then Monty can open either no. 2 or no. 3 to display a goat.

If the contestant picks Door no. 1 and the car is behind no. 2, then Monty opens no. 3.

If the contestant picks Door no. 1 and the car is behind no. 3, then Monty opens no. 2.

By switching after a door is opened, the contestant gets the benefit of choosing two doors rather than one. I will try to persuade you in three different ways that this analysis is correct.

The first is empirical. In 2008, *New York Times* columnist John Tierney wrote about the Monty Hall phenomenon.[1] The *Times* then constructed an interactive feature that allows you to play the game yourself, including the decision to switch or not. (There are even little goats and cars that pop out from behind the doors.) The game keeps track of your success when you switch doors after making your initial decision compared with when you do not. Try it yourself.* I paid one of my children to play the game 100 times, switching each time. I paid her brother to play the game 100 times without switching. The switcher won 72 times; the nonswitcher won 33 times. Both received two dollars for their efforts.

The data from episodes of *Let's Make a Deal* suggest the same thing. According to Leonard Mlodinow, author of *The Drunkard's Walk*, those contestants who switched their choice won about twice as often as those who did not.[2]

My second explanation gets at the intuition. Let's suppose the rules were modified slightly. Assume that the contestant begins by picking one of the three doors: no. 1, no. 2, or no. 3, just as the game is ordinarily played. But then, before any door is opened to reveal a goat, Monty says, "Would you like to give up your choice in exchange for *both of the other doors that you did not choose*?" So if you picked Door no. 1, you could ditch that door in exchange for what is behind no. 2 and no. 3. If you picked no. 3, you could switch to no. 1 and no. 2. And so on.

That would not be a particularly hard decision. Obviously you should give up one door in exchange for two, as it increases your chances of winning from $1/3$ to $2/3$. Here is the intriguing part: *That is exactly what Monty Hall allows you to do in the real game after he reveals the goat.* The fundamental insight is that if you were to choose two doors, one of them would always have a goat behind it anyway. When he opens a door to reveal a goat before asking if you'd like to switch, he's doing you a huge favor! He's saying (in effect), "There is a two-thirds chance that the car is behind one of the doors you didn't choose, and look, it's not that one!"

Think of it this way. Suppose you picked Door no. 1. Monty then

---

* You can play the game at http://www.nytimes.com/2008/04/08/science/08monty.html?_r=2&oref=slogin&oref=slogin.

offers you the option to take Doors 2 and 3 instead. You take the offer, giving up one door and getting two, meaning that you can reasonably expect to win the car $2/3$ of the time. At that point, what if Monty were to open Door no. 3—one of your doors—to reveal a goat? Should you feel less certain about your decision? Of course not. If the car were behind no. 3, he would have opened no. 2! *He's shown you nothing.*

When the game is played normally, Monty is really giving you a choice between the door you originally picked and the other two doors, only one of which could possibly have a car behind it. When he opens a door to reveal a goat, he's merely doing you the courtesy of showing you which of the other two doors does not have the car. You have the same probability of winning in both of the following scenarios:

1. Choosing Door no. 1, then agreeing to switch to Door no. 2 *and* Door no. 3 before any door is opened.
2. Choosing Door no. 1, then agreeing to switch to Door no. 2 after Monty reveals a goat behind Door no. 3 (or choosing no. 3 after he reveals a goat behind no. 2).

In both cases, switching gives you the benefit of two doors instead of one, and you can therefore double your chances of winning, from $1/3$ to $2/3$.

My third explanation is a more extreme version of the same basic intuition. Assume that Monty Hall offers you a choice from among 100 doors rather than just three. After you pick your door, say, no. 47, he opens 98 other doors with goats behind them. Now there are only two doors that remain closed, no. 47 (your original choice) and one other, say, no. 61. Should you switch?

Of course you should. There is a 99 percent chance that the car was behind one of the doors that you did not originally choose. Monty did you the favor of opening 98 of those doors that you did not choose, all of which he knew did not have the car behind them. There is only a 1 in 100 chance that your original pick was correct (no. 47). There is a 99 in 100 chance that your original pick was not correct. And if your original pick was not correct, then the car is sitting behind the other door, no. 61. If you want to win 99 times out of 100, you should switch to no. 61.

In short, if you ever find yourself as a contestant on *Let's Make a Deal*, you should definitely switch doors when Monty Hall (or his replacement) gives you the option. The more broadly applicable lesson is that your gut instinct on probability can sometimes steer you astray.

CHAPTER 6

Problems with Probability

How overconfident math geeks nearly destroyed the global financial system

Statistics cannot be any smarter than the people who use them. And in some cases, they can make smart people do dumb things. One of the most irresponsible uses of statistics in recent memory involved the mechanism for gauging risk on Wall Street prior to the 2008 financial crisis. At that time, firms throughout the financial industry used a common barometer of risk, the Value at Risk model, or VaR. In theory, VaR combined the elegance of an indicator (collapsing lots of information into a single number) with the power of probability (attaching an expected gain or loss to each of the firm's assets or trading positions). The model assumed that there is a range of possible outcomes for every one of the firm's investments. For example, if the firm owns General Electric stock, the value of those shares can go up or down. When the VaR is being calculated for some short period of time, say, one week, the most likely outcome is that the shares will have roughly the same value at the end of that stretch as they had at the beginning. There is a smaller chance that the shares may rise or fall by 10 percent. And an even smaller chance that they may rise or fall 25 percent, and so on.

On the basis of past data for market movements, the firm's quantitative experts (often called "quants" in the industry and "rich nerds" everywhere else) could assign a dollar figure, say $13 million, that represented the maximum that the firm could lose on that position over the time period being examined, with 99 percent probability. In other words, 99 times out of 100 the firm would not lose more than $13 million on a particular trading position; 1 time out of 100, it would.

Remember that last part, because it will soon become important.

Prior to the financial crisis of 2008, firms trusted the VaR model to quantify their overall risk. If a single trader had 923 different open positions (investments that could move up or down in value), each of those investments could be evaluated as described above for the General Electric stock; from there, the trader's total portfolio risk could be calculated. The formula even took into account the correlations among different positions. For example, if two investments had expected returns that were negatively correlated, a loss in one would likely have been offset by a gain in the other, making the two investments together less risky than either one separately. Overall, the head of the trading desk would know that bond trader Bob Smith has a 24-hour VaR (the value at risk over the next 24 hours) of $19 million, again with 99 percent probability. The most that Bob Smith could lose over the next 24 hours would be $19 million, 99 times out of 100.

Then, even better, the aggregate risk for the firm could be calculated at any point in time by taking the same basic process one step further. The underlying mathematical mechanics are obviously fabulously complicated, as firms had a dizzying array of investments in different currencies, with different amounts of leverage (the amount of money that was borrowed to make the investment), trading in markets with different degrees of liquidity, and so on. Despite all that, the firm's managers ostensibly had a precise measure of the magnitude of the risk that the firm had taken on at any moment in time. As former *New York Times* business writer Joe Nocera has explained, "VaR's great appeal, and its great selling point to people who do not happen to be quants, is that it expresses risk as a single number, a dollar figure, no less."[1] At J. P. Morgan, where the VaR model was developed and refined, the daily VaR calculation was known as the "4:15 report" because it would be on the desks of top executives every afternoon at 4:15, just after the American financial markets had closed for the day.

Presumably this was a good thing, as more information is generally better, particularly when it comes to risk. After all, probability is a powerful tool. Isn't this just the same kind of calculation that the Schlitz executives did before spending a lot of money on blind taste tests at half-time of the Super Bowl?

Not necessarily. VaR has been called "potentially catastrophic," "a fraud," and many other things not fit for a family book about statistics like this one. In particular, the model has been blamed for the onset and severity of the financial crisis. The primary critique of VaR is that the underlying risks associated with financial markets are not as predictable as a coin flip or even a blind taste test between two beers. The false precision embedded in the models created a false sense of security. The VaR was like a faulty speedometer, which is arguably worse than no speedometer at all. If you place too much faith in the broken speedometer, you will be oblivious to other signs that your speed is unsafe. In contrast, if there is no speedometer at all, you have no choice but to look around for clues as to how fast you are really going.

By around 2005, with the VaR dropping on desks at 4:15 every weekday, Wall Street was driving pretty darn fast. Unfortunately, there were two huge problems with the risk profiles encapsulated by the VaR models. First, the underlying probabilities on which the models were built were based on past market movements; however, in financial markets (unlike beer tasting), the future does not necessarily look like the past. There was no intellectual justification for assuming that the market movements from 1980 to 2005 were the best predictor of market movements after 2005. In some ways, this failure of imagination resembles the military's periodic mistaken assumption that the next war will look like the last one. In the 1990s and early 2000s, commercial banks were using lending models for home mortgages that assigned zero probability to large declines in housing prices.[2] Housing prices had never before fallen as far and as fast as they did beginning in 2007. But that's what happened. Former Federal Reserve chairman Alan Greenspan explained to a congressional committee after the fact, "The whole intellectual edifice, however, collapsed in the summer of [2007] because the data input into the risk management models generally covered only the past two decades, a period of euphoria. Had instead the models been fitted more appropriately to historic periods of stress, capital requirements would have been much higher and the financial world would be in far better shape, in my judgment."[3]

Second, even if the underlying data could accurately predict future risk, the 99 percent assurance offered by the VaR model was dangerously useless, *because it's the 1 percent that is going to really mess you up*. Hedge fund manager David Einhorn explained, "This is like an air bag that works all the time, except when you have a car accident." If a firm has a Value at Risk of $500 million, that can be interpreted to mean that the firm has a 99 percent chance of losing no more than $500 million over the time period specified. Well, hello, that also means that the firm has a 1 percent chance of losing more than $500 million—much, much more under some circumstances. In fact, the models had nothing to say about how bad that 1 percent scenario might turn out to be. Very little attention was devoted to the "tail risk," the small risk (named for the tail of the distribution) of some catastrophic outcome. (If you drive home from a bar with a blood alcohol level of .15, there is probably less than a 1 percent chance that you will crash and die; that does not make it a sensible thing to do.) Many firms compounded this error by making unrealistic assumptions about their preparedness for rare events. Former treasury secretary Hank Paulson has explained that many firms assumed they could raise cash in a pinch by selling assets.[4] But during a crisis, every other firm needs cash, too, so all are trying to sell the same kinds of assets. It's the risk management equivalent of saying, "I don't need to stock up on water because if there is a natural disaster, I'll just go to the supermarket and buy some." Of course, after an asteroid hits your town, fifty thousand other people are also trying to buy water; by the time you get to the supermarket, the windows are broken and the shelves are empty.

The fact that you've never contemplated that your town might be flattened by a massive asteroid was exactly the problem with VaR. Here is *New York Times* columnist Joe Nocera again, summarizing thoughts of Nicholas Taleb, author of *The Black Swan: The Impact of the Highly Improbable* and a scathing critic of VaR: "The greatest risks are never the ones you can see and measure, but the ones you can't see and therefore can never measure. The ones that seem so far outside the boundary of normal probability that you can't imagine they could happen in your lifetime—even though, of course, they do happen, more often than you care to realize."

In some ways, the VaR debacle is the opposite of the Schlitz example in Chapter 5. Schlitz was operating with a known probability distribution. Whatever data the company had on the likelihood of blind taste testers' choosing Schlitz was a good estimate of how similar testers would behave live at halftime. Schlitz even managed its downside by performing the whole test on men who said they liked the other beers better. Even if no more than twenty-five Michelob drinkers chose Schlitz (an almost impossibly low outcome), Schlitz could still claim that one in four beer drinkers ought to consider switching. Perhaps most important, this was all just beer, not the global financial system. The Wall Street quants made three fundamental errors. First, they confused precision with accuracy. The VaR models were just like my golf range finder when it was set to meters instead of yards: exact and wrong. The false precision led Wall Street executives to believe that they had risk on a leash when in fact they did not. Second, the estimates of the underlying probabilities were wrong. As Alan Greenspan pointed out in testimony quoted earlier in the chapter, the relatively tranquil and prosperous decades before 2005 should not have been used to create probability distributions for what might happen in the markets in the ensuing decades. This is the equivalent of walking into a casino and thinking that you will win at roulette 62 percent of the time because that's what happened last time you went gambling. It would be a long, expensive evening. Third, firms neglected their "tail risk." The VaR models predicted what would happen 99 times out of 100. That's the way probability works (as the second half of the book will emphasize repeatedly). Unlikely things happen. In fact, over a long enough period of time, they are not even that unlikely. People get hit by lightning all the time. My mother has had three holes in one.

The statistical hubris at commercial banks and on Wall Street ultimately contributed to the most severe global financial contraction since the Great Depression. The crisis that began in 2008 destroyed trillions of dollars in wealth in the United States, drove unemployment over 10 percent, created waves of home foreclosures and business failures, and saddled governments around the world with huge debts as they struggled to contain the economic damage. This is a sadly ironic outcome, given that sophisticated tools like VaR were designed to mitigate risk.

Probability offers a powerful and useful set of tools—many of which can be employed correctly to understand the world or incorrectly to wreak havoc on it. In sticking with the "statistics as a powerful weapon" metaphor that I've used throughout the book, I will paraphrase the gun rights lobby: Probability doesn't make mistakes; people using probability make mistakes. The balance of this chapter will catalog some of the most common probability-related errors, misunderstandings, and ethical dilemmas.

Assuming events are independent when they are not. The probability of flipping heads with a fair coin is ½. The probability of flipping two heads in a row is $(1/2)^2$, or ¼, since the likelihood of two independent events' *both* happening is the product of their individual probabilities. Now that you are armed with this powerful knowledge, let's assume that you have been promoted to head of risk management at a major airline. Your assistant informs you that the probability of a jet engine's failing for any reason during a transatlantic flight is 1 in 100,000. Given the number of transatlantic flights, this is not an acceptable risk. Fortunately each jet making such a trip has at least two engines. Your assistant has calculated that the risk of both engines' shutting down over the Atlantic must be $(1/100{,}000)^2$, or 1 in 10 billion, which is a reasonable safety risk. This would be a good time to tell your assistant to use up his vacation days before he is fired. The two engine failures are not independent events. If a plane flies through a flock of geese while taking off, both engines are likely to be compromised in a similar way. The same would be true of many other factors that affect the performance of a jet engine, from weather to improper maintenance. If one engine fails, the probability that the second engine fails is going to be significantly higher than 1 in 100,000.

Does this seem obvious? It was not obvious throughout the 1990s as British prosecutors committed a grave miscarriage of justice because of an improper use of probability. As with the hypothetical jet engine example, the statistical mistake was in assuming that several events were independent (as in flipping a coin) rather than dependent (when a certain outcome makes a similar outcome more likely in the future). This mistake was real, however, and innocent people were sent to jail as a result.

The mistake arose in the context of sudden infant death syndrome (SIDS), a phenomenon in which a perfectly healthy infant dies in his or her crib. (The Brits refer to SIDS as a "cot death.") SIDS was a medical mystery that attracted more attention as infant deaths from other causes became less common.* Because these infant deaths were so mysterious and poorly understood, they bred suspicion. Sometimes that suspicion was warranted. SIDS was used on occasion to cover up parental negligence or abuse; a postmortem exam cannot necessarily distinguish natural deaths from those in which foul play is involved. British prosecutors and courts became convinced that one way to separate foul play from natural deaths would be to focus on families in which there were multiple cot deaths. Sir Roy Meadow, a prominent British pediatrician, was a frequent expert witness on this point. As the British news magazine the *Economist* explains, "What became known as Meadow's Law—the idea that one infant death is a tragedy, two are suspicious and three are murder—is based on the notion that if an event is rare, two or more instances of it in the same family are so improbable that they are unlikely to be the result of chance."[5] Sir Meadow explained to juries that the chance that a family could have two infants die suddenly of natural causes was an extraordinary 1 in 73 million. He explained the calculation: Since the incidence of a cot death is rare, 1 in 8,500, the chance of having two cot deaths in the same family would be $(1/8{,}500)^2$ which is roughly 1 in 73 million. This reeks of foul play. That's what juries decided, sending many parents to prison on the basis of this testimony on the statistics of cot deaths (often without any corroborating medical evidence of abuse or neglect). In some cases, infants were taken away from their parents at birth because of the unexplained death of a sibling.

---

* SIDS is still a medical mystery, though many of the risk factors have been identified. For example, infant deaths can be reduced sharply by putting babies to sleep on their backs.

The *Economist* explained how a misunderstanding of statistical independence became a flaw in the Meadow testimony:

> There is an obvious flaw in this reasoning, as the Royal Statistical Society, protective of its derided subject, has pointed out. The probability calculation works fine, so long as it is certain that cot deaths are entirely random and not linked by some unknown factor. But with something as mysterious as cot deaths, it is quite possible that there is a link—something genetic, for instance, which would make a family that had suffered one cot death more, not less, likely to suffer another. And since those women were convicted, scientists have been suggesting that there may be just such a link.

In 2004, the British government announced that it would review 258 trials in which parents had been convicted of murdering their infant children.

Not understanding when events ARE independent. A different kind of mistake occurs when events that *are* independent are not treated as such. If you find yourself in a casino (a place, statistically speaking, that you should not go to), you will see people looking longingly at the dice or cards and declaring that they are "due." If the roulette ball has landed on black five times in a row, then clearly now it must turn up red. No, no, no! The probability of the ball's landing on a red number remains unchanged: 16/38. The belief otherwise is sometimes called "the gambler's fallacy." In fact, if you flip a fair coin 1,000,000 times and get 1,000,000 heads in a row, the probability of getting tails on the next flip is still ½. The very definition of statistical independence between two events is that the outcome of one has no effect on the outcome of the other. Even if you don't find the statistics persuasive, you might ask yourself about the physics: How can flipping a series of tails in a row make it more likely that the coin will turn up heads on the next flip?

Even in sports, the notion of streaks may be illusory. One of the most famous and interesting probability-related academic papers refutes the common notion that basketball players periodically develop a streak of good shooting during a game, or "a hot hand." Certainly most sports fans would tell you that a player who makes a shot is more likely to hit the next shot than a player who has just missed. Not according to research by Thomas Gilovich, Robert Vallone, and Amos Tversky, who tested the hot hand in three different ways.[6] First, they analyzed shooting data for the Philadelphia 76ers home games during the 1980–81 season. (At the time, similar data were not available for other teams in the NBA.) They found "no evidence for a positive correlation between the outcomes of successive shots." Second, they did the same thing for free throw data for the Boston Celtics, which produced the same result. And last, they did a controlled experiment with members of the Cornell men's and women's basketball teams. The players hit an average of 48 percent of their field goals after hitting their last shot and 47 percent after missing. For fourteen of twenty-six players, the correlation between making one shot and then making the next was negative. Only one player showed a significant positive correlation between one shot and the next.

That's not what most basketball fans will tell you. For example, 91 percent of basketball fans surveyed at Stanford and Cornell by the authors of the paper agreed with the statement that a player has a better chance of making his next shot after making his last two or three shots than he does after missing his last two or three shots. The significance of the "hot hand" paper lies in the difference between the perception and the empirical reality. The authors note, "People's intuitive conceptions of randomness depart systematically from the laws of chance." We see patterns where none may really exist.

Like cancer clusters.

Clusters happen. You've probably read the story in the newspaper, or perhaps seen the news exposé: Some statistically unlikely number of people in a particular area have contracted a rare form of cancer. It must be the water, or the local power plant, or the cell phone tower. Of course, any one of those things might really be causing adverse health outcomes. (Later chapters will explore how statistics can identify such causal relationships.) But this cluster of cases may also be the product of pure chance, even when the number of cases appears highly improbable. Yes, the probability that five people in the same school or church or workplace will contract the same rare form of leukemia may be one in a million, *but there are millions of schools and churches and workplaces*. It's not highly improbable that five people might get the same rare form of leukemia in one of those places. We just aren't thinking about all the schools and churches and workplaces where this hasn't happened. To use a different variation on the same basic example, the chance of winning the lotto may be 1 in 20 million, but none of us is surprised when *someone* wins, because millions of tickets have been sold. (Despite my general aversion to lotteries, I do admire the Illinois slogan: "Someone's gonna Lotto, might as well be you.")

Here is an exercise that I do with my students to make the same basic point. The larger the class, the better it works. I ask everyone in the class to take out a coin and stand up. We all flip the coin; anyone who flips heads must sit down. Assuming we start with 100 students, roughly 50 will sit down after the first flip. Then we do it again, after which 25 or so are still standing. And so on. More often than not, there will be a student standing at the end who has flipped five or six tails in a row. At that point, I ask the student questions like "How did you do it?" and "What are the best training exercises for flipping so many tails in a row?" or "Is there a special diet that helped you pull off this impressive accomplishment?" These questions elicit laughter because the class has just watched the whole process unfold; they know that the student who flipped six tails in a row has no special coin-flipping talent. He or she just happened to be the one who ended up with a lot of tails. When we see an anomalous event like that out of context, however, we assume that something besides randomness must be responsible.

The prosecutor's fallacy. Suppose you hear testimony in court to the following effect: (1) a DNA sample found at the scene of a crime matches a sample taken from the defendant; and (2) there is only one chance in a million that the sample recovered at the scene of the crime would match anyone's besides the defendant. (For the sake of this example, you can assume that the prosecution's probabilities are correct.) On the basis of that evidence, would you vote to convict?

I sure hope not.

The prosecutor's fallacy occurs when the context surrounding statistical evidence is neglected. Here are two scenarios, each of which could explain the DNA evidence being used to prosecute the defendant.

Defendant 1: This defendant, a spurned lover of the victim, was arrested three blocks from the crime scene carrying the murder weapon. After he was arrested, the court compelled him to offer a DNA sample, which matched a sample taken from a hair found at the scene of the crime.

Defendant 2: This defendant was convicted of a similar crime in a different state several years ago. As a result of that conviction, his DNA was included in a national DNA database of over a million violent felons. The DNA sample taken from the hair found at the scene of the crime was run through that database and matched to this individual, who has no known association with the victim.

As noted above, in both cases the prosecutor can rightfully say that the DNA sample taken from the crime scene matches the defendant's and that there is only a one in a million chance that it would match with anyone else's. But in the case of Defendant 2, there is a darn good chance that he could be that random someone else, the one in a million guy whose DNA just happens to be similar to the real killer's by chance. *Because the chances of finding a coincidental one in a million match are relatively high if you run the sample through a database with samples from a million people.*

Reversion to the mean (or regression to the mean). Perhaps you've heard of the *Sports Illustrated* jinx, whereby individual athletes or teams featured on the cover of *Sports Illustrated* subsequently see their performance fall off. One explanation is that being on the cover of the magazine has some adverse effect on subsequent performance. The more statistically sound explanation is that teams and athletes appear on its cover after some anomalously good stretch (such as a twenty-game winning streak) and that their subsequent performance merely reverts back to what is normal, or the mean. This is the phenomenon known as reversion to the mean. Probability tells us that any outlier—an observation that is particularly far from the mean in one direction or the other—is likely to be followed by outcomes that are more consistent with the long-term average.

Reversion to the mean can explain why the Chicago Cubs always seem to pay huge salaries for free agents who subsequently disappoint fans like me. Players are able to negotiate huge salaries with the Cubs after an exceptional season or two. Putting on a Cubs uniform does not necessarily make these players worse (though I would not necessarily rule that out); rather, the Cubs pay big bucks for these superstars at the end of some exceptional stretch—an outlier year or two—after which their performance for the Cubs reverts to something closer to normal.

The same phenomenon can explain why students who do much better than they normally do on some kind of test will, on average, do slightly worse on a retest, and students who have done worse than usual will tend to do slightly better when retested. One way to think about this mean reversion is that performance—both mental and physical—consists of some underlying talent-related effort plus an element of luck, good or bad. (Statisticians would call this random error.) In any case, those individuals who perform far above the mean for some stretch are likely to have had luck on their side; those who perform far below the mean are likely to have had bad luck. (In the case of an exam, think about students guessing right or wrong; in the case of a baseball player, think about a hit that can either go foul or land one foot fair for a triple.) When a spell of very good luck or very bad luck ends—as it inevitably will—the resulting performance will be closer to the mean.

Imagine that I am trying to assemble a superstar coin-flipping team (under the erroneous impression that talent matters when it comes to coin flipping). After I observe a student flipping six tails in a row, I offer him a ten-year, $50 million contract. Needless to say, I'm going to be disappointed when this student flips only 50 percent tails over those ten years.

At first glance, reversion to the mean may appear to be at odds with the "gambler's fallacy." After the student throws six tails in a row, is he "due" to throw heads or not? The probability that he throws heads on the next flip is the same as it always is: ½. The fact that he has thrown lots of tails in a row does not make heads more likely on the next flip. Each flip is an independent event. However, we can expect the results of the ensuing flips to be consistent with what probability predicts, which is half heads and half tails, rather than what it has been in the past, which is all tails. It's a virtual certainty that someone who has flipped all tails will begin throwing more heads in the ensuing 10, 20, or 100 flips. And the more flips, the more closely the outcome will resemble the 50-50 mean outcome that the law of large numbers predicts. (Or, alternatively, we should start looking for evidence of fraud.)

As a curious side note, researchers have also documented a *Businessweek* phenomenon. When CEOs receive high-profile awards, including being named one of *Businessweek*'s "Best Managers," their companies subsequently underperform over the next three years as measured by both accounting profits and stock price. However, unlike the *Sports Illustrated* effect, this effect appears to be more than reversion to the mean. According to Ulrike Malmendier and Geoffrey Tate, economists at the University of California at Berkeley and UCLA, respectively, when CEOs achieve "superstar" status, they get distracted by their new prominence.[7] They write their memoirs. They are invited to sit on outside boards. They begin searching for trophy spouses. (The authors propose only the first two explanations, but I find the last one plausible as well.) Malmendier and Tate write, "Our results suggest that media-induced superstar culture leads to behavioral distortions beyond mere mean reversion." In other words, when a CEO appears on the cover of *Businessweek*, sell the stock.

Statistical discrimination. When is it okay to act on the basis of what probability tells us is likely to happen, and when is it not okay? In 2003, Anna Diamantopoulou, the European commissioner for employment and social affairs, proposed a directive declaring that insurance companies may not charge different rates to men and women, because it violates the European Union's principle of equal treatment.[8] To insurers, however, gender-based premiums aren't discrimination; they're just statistics. Men typically pay more for auto insurance because they crash more. Women pay more for annuities (a financial product that pays a fixed monthly or yearly sum until death) because they live longer. Obviously many women crash more than many men, and many men live longer than many women. But, as explained in the last chapter, insurance companies don't care about that. They care only about what happens on average, because if they get that right, the firm will make money. The interesting thing about the European Commission policy banning gender-based insurance premiums, which is being implemented in 2012, is that the authorities are not pretending that gender is unrelated to the risks being insured; they are simply declaring that disparate rates based on sex are unacceptable.*

At first, that feels like an annoying nod to political correctness. Upon reflection, I'm not so sure. Remember all that impressive stuff about preventing crimes before they happen? Probability can lead us to some intriguing but distressing places in this regard. How should we react when our probability-based models tell us that methamphetamine smugglers from Mexico are most likely to be Hispanic men aged between eighteen and thirty and driving red pickup trucks between 9:00 p.m. and midnight when we also know that the vast majority of Hispanic men who fit that profile are not smuggling methamphetamine? Yep, I used the profiling word, because that's the less glamorous description of the predictive analytics that I described so glowingly in the last chapter, or at least one potential aspect of it.

Probability tells us what is more likely and what is less likely. Yes, that is just basic statistics—the tools described over the last few chapters. But it is also statistics with social implications. If we want to catch violent criminals and terrorists and drug smugglers and other individuals with the potential to do enormous harm, then we ought to use every tool at our disposal. Probability can be one of those tools. It would be naïve to think that gender, age, race, ethnicity, religion, and country of origin collectively tell us nothing about anything related to law enforcement.

But what we can or should do with that kind of information (assuming it has some predictive value) is a philosophical and legal question, not a statistical one. We're getting more and more information every day about more and more things. Is it okay to discriminate if the data tell us that we'll be right far more often than we're wrong? (This is the origin of the term "statistical discrimination," or "rational discrimination.") The

* The policy change was ultimately precipitated by a 2011 ruling by the Court of Justice of the European Union that different premiums for men and women constitute sex discrimination.

same kind of analysis that can be used to determine that people who buy birdseed are less likely to default on their credit cards (yes, that's really true) can be applied everywhere else in life. How much of that is acceptable? If we can build a model that identifies drug smugglers correctly 80 out of 100 times, what happens to the poor souls in the 20 percent—*because our model is going to harass them over and over and over again*.

The broader point here is that our ability to analyze data has grown far more sophisticated than our thinking about what we ought to do with the results. You can agree or disagree with the European Commission decision to ban gender-based insurance premiums, but I promise you it will not be the last tricky decision of that sort. We like to think of numbers as "cold, hard facts." If we do the calculations right, then we must have *the right answer*. The more interesting and dangerous reality is that we can sometimes do the calculations correctly and end up blundering in a dangerous direction. We can blow up the financial system or harass a twenty-two-year-old white guy standing on a particular street corner at a particular time of day, because, according to our statistical model, he is almost certainly there to buy drugs. For all the elegance and precision of probability, there is no substitute for thinking about what calculations we are doing and why we are doing them.

CHAPTER 7

The Importance of Data

"Garbage in, garbage out"

In the spring of 2012, researchers published a striking finding in the esteemed journal *Science*. According to this cutting-edge research, when male fruit flies are spurned repeatedly by female fruit flies, they drown their sorrows in alcohol. The *New York Times* described the study in a front page article: "They were young males on the make, and they struck out not once, not twice, but a dozen times with a group of attractive females hovering nearby. So they did what so many men do after being repeatedly rejected: they got drunk, using alcohol as a balm for unfulfilled desire."[1]

This research advances our understanding of the brain's reward system, which in turn can help us find new strategies for dealing with drug and alcohol dependence. A substance abuse expert described reading the study as "looking back in time, to see the very origins of the reward circuit that drives fundamental behaviors like sex, eating and sleeping."

Since I am not an expert in this field, I had two slightly different reactions upon reading about spurned fruit flies. First, it made me nostalgic for college. Second, my inner researcher got to wondering how fruit flies get drunk. Is there a miniature fruit fly bar, with assorted fruit based liquors and an empathetic fruit fly bartender? Is country western music playing in the background? Do fruit flies even like country western music?

It turns out that the design of the experiment was devilishly simple. One group of male fruit flies was allowed to mate freely with virgin females. Another group of males was released among female fruit flies that had already mated and were therefore indifferent to the males' amorous overtures. Both sets of male fruit flies were then offered feeding straws that offered a choice between standard fruit fly fare, yeast and sugar, and the "hard stuff": yeast, sugar, and 15 percent alcohol. The males who had spent days trying to mate with indifferent females were significantly more likely to hit the booze.

The levity notwithstanding, these results have important implications for humans. They suggest a connection between stress, chemical responses in the brain, and an appetite for alcohol. However, the results are not a triumph of statistics. They are a triumph of data, which made relatively basic statistical analysis possible. The genius of this study was figuring out a way to create a group of sexually satiated male fruit flies and a group of sexually frustrated male fruit flies—and then to find a way to compare their drinking habits. Once the researchers did that, the number crunching wasn't any more complicated than that of a typical high school science fair project.

Data are to statistics what a good offensive line is to a star quarterback. In front of every star quarterback is a good group of blockers. They usually don't get much credit. But without them, you won't ever see a star quarterback. Most statistics books assume that you are using good data, just as a cookbook assumes that you are not buying rancid meat and rotten vegetables. But even the finest recipe isn't going to salvage a meal that begins with spoiled ingredients. So it is with statistics; no amount of fancy analysis can make up for fundamentally flawed data. Hence the expression "garbage in, garbage out." Data deserve respect, just like offensive linemen.

We generally ask our data to do one of three things. First, we may demand a data sample that is representative of some larger group or population. If we are trying to gauge voters' attitudes toward a particular political candidate, we will need to interview a sample of prospective voters who are representative of all voters in the relevant political jurisdiction. (And remember, we don't want a sample that is representative of everyone *living* in that jurisdiction; we want a sample of those *who are likely to vote.*) One of the most powerful findings in statistics, which will be explained in greater depth over the next two chapters, is that inferences made from reasonably large, properly drawn samples can be every bit as accurate as attempting to elicit the same information from the entire population.

The easiest way to gather a representative sample of a larger population is to select some subset of that population randomly. (Shockingly, this is known as a simple random sample.) The key to this methodology is that each observation in the relevant population must have an equal chance of being included in the sample. If you plan to survey a random sample of 100 adults in a neighborhood with 4,328 adult residents, your methodology has to ensure that each of those 4,328 residents has the same probability of ending up as one of the 100 adults who are surveyed. Statistics books almost always illustrate this point by drawing colored marbles out of an urn. (In fact, it's about the only place where one sees the word "urn" used with any regularity.) If there are 60,000 blue marbles and 40,000 red marbles in a giant urn, then the most likely composition of a sample of 100 marbles drawn randomly from the urn would be 60 blue marbles and 40 red marbles. If we did this more than once, there would obviously be deviations from sample to sample—some might have 62 blue marbles and 38 red marbles, or 58 blue and 42 red. But the chances of drawing any random sample that deviates hugely from the composition of marbles in the urn are very, very low.

Now, admittedly, there are some practical challenges here. Most populations we care about tend to be more complicated than an urn full of marbles. How, exactly, would one select a random sample of the American adult population to be included in a telephone poll? Even a seemingly elegant solution like a telephone random dialer has potential flaws. Some individuals (particularly low-income persons) may not have a telephone. Others (particularly high-income persons) may be more prone to screen calls and choose not to answer. Chapter 10 will outline some of the strategies that polling firms use to surmount these kinds of sampling challenges (most of which got even more complicated with the advent of cell phones). The key idea is that a properly drawn sample will look like the population from which it is drawn. In terms of intuition, you can envision sampling a pot of soup with a single spoonful. If you've stirred your soup adequately, a single spoonful can tell you how the whole pot tastes.

A statistics text will include far more detail on sampling methods. Polling firms and market research companies spend their days figuring out how to get good representative data from various populations in the most cost-effective way. For now, you should appreciate several important things: (1) A representative sample is a fabulously important thing, for it opens the door to some of the most powerful tools that statistics has to offer. (2) Getting a good sample is harder than it looks. (3) Many of the most egregious statistical assertions are caused by good statistical methods applied to bad samples, not the opposite. (4) Size matters, and bigger is better. The details will be explained in the coming chapters, but it should be intuitive that a larger sample will help to smooth away any freak variation. (A bowl of soup will be an even better test than a spoonful.) One crucial caveat is that a bigger sample will not make up for errors in its composition, or "bias." A bad sample is a bad sample. No supercomputer or fancy formula is going to rescue the validity of your *national* presidential poll if the respondents are drawn only from a telephone survey of Washington, D.C., residents. The residents of Washington, D.C., don't vote like the rest of America; calling 100,000 D.C. residents rather than 1,000 is not going to fix that fundamental problem with your poll. In fact, a large, biased sample is arguably worse than a small, biased sample because it will give a false sense of confidence regarding the results.

The second thing we often ask of data is that they provide some source of comparison. Is a new medicine more effective than the current treatment? Are ex-convicts who receive job training less likely to return to prison than ex-convicts who do not receive such training? Do students who attend charter schools perform better than similar students who attend regular public schools?

In these cases, the goal is to find two groups of subjects who are broadly similar except for the application of whatever "treatment" we care about. In a social science context, the word "treatment" is broad enough to encompass anything from being a sexually frustrated fruit fly to receiving an income tax rebate. As with any other application of the scientific method, we are trying to isolate the impact of *one specific intervention or attribute*. This was the genius of the fruit fly experiment. The researchers figured out a way to create a control group (the males who mated) and a "treatment" group (the males who were shot down); the subsequent difference in their drinking behaviors can then be attributed to whether they were sexually spurned or not.

In the physical and biological sciences, creating treatment and control groups is relatively straightforward. Chemists can make small variations from test tube to test tube and then study the difference in outcomes. Biologists can do the same thing with their petri dishes. Even most animal testing is simpler than trying to get fruit flies to drink alcohol. We can have one group of rats exercise regularly on a treadmill and then compare their mental acuity in a maze with the performance of another group of rats that didn't exercise. But when humans become involved, things grow more complicated. Sound statistical analysis often requires a treatment and a control group, yet we cannot force people to do the things that we make laboratory rats do. (And many people do not like making even the lab rats do these things.) Do repeated concussions cause serious neurological problems later in life? This is a really important question. The future of football (and perhaps other sports) hangs on the answer. Yet it is a question that cannot be answered with experiments on humans. So unless and until we can teach fruit flies to wear helmets and run the spread offense, we have to find other ways to study the long-term impact of head trauma.

One recurring research challenge with human subjects is creating treatment and control groups that differ *only* in that one group is getting the treatment and the other is not. For this reason, the "gold standard" of research is randomization, a process by which human subjects (or schools, or hospitals, or whatever we're studying) are randomly assigned to either the treatment or the control group. We do not assume that all the experimental subjects are identical. Instead, probability becomes our friend (once again), and we assume that randomization will evenly divide all relevant characteristics between the two groups—both the characteristics we can observe, like race or income, but also confounding characteristics that we cannot measure or had not considered, such as perseverance or faith.

The third reason we collect data is, to quote my teenage daughter, "Just because." We sometimes have no specific idea what we will do with the information—but we suspect it will come in handy at some point. This is similar to a crime scene detective who demands that all possible evidence be captured so that it can be sorted later for clues. Some of this evidence will prove useful, some will not. If we knew exactly what would be useful, we probably would not need to be doing the investigation in the first place.

You probably know that smoking and obesity are risk factors for heart disease. You probably don't know that a long-running study of the residents of Framingham, Massachusetts, helped to clarify those relationships. Framingham is a suburban town of some 67,000 people about twenty miles west of Boston. To nonresearchers, it is best known as a suburb of Boston with reasonably priced housing and convenient access to the impressive and upscale Natick Mall. To researchers, Framingham is best known as the home of the Framingham Heart Study, one of the most successful and influential longitudinal studies in the history of modern science.

A longitudinal study collects information on a large group of subjects at many different points in time, such as once every two years. The same participants may be interviewed periodically for ten, twenty, or even fifty years after they enter the study, creating a remarkably rich trove of information. In the case of the Framingham study, researchers gathered information on 5,209 adult residents of Framingham in 1948: height, weight, blood pressure, educational background, family structure, diet, smoking behavior, drug use, and so on. Most important, researchers have gathered follow-up data *from the same participants* ever since (and also data on their offspring, to examine genetic factors related to heart disease). The Framingham data have been used to produce over two thousand academic articles since 1950, including nearly a thousand between 2000 and 2009.

These studies have produced findings crucial to our understanding of cardiovascular disease, many of which we now take for granted: cigarette smoking increases the risk of heart disease (1960); physical activity reduces the risk of heart disease and obesity increases it (1967); high blood pressure increases the risk of stroke (1970); high levels of HDL cholesterol (henceforth known as the "good cholesterol") reduce the risk of death (1988); individuals with parents and siblings who have cardiovascular disease are at significantly higher risk of the same (2004 and 2005).

Longitudinal data sets are the research equivalent of a Ferrari. The data are particularly valuable when it comes to exploring causal relationships that may take years or decades to unfold. For example, the Perry Preschool Study began in the late 1960s with a group of 123 African American three- and four-year-olds from poor families. The participating children were randomly assigned into a group that received an intensive preschool program and a comparison group that did not. Researchers then measured various outcomes for both groups *for the next forty years*. The results make a compelling case for the benefits of early childhood education. The students who received the intensive preschool experience had higher IQs at age five. They were more likely to graduate from high school. They had higher earnings at age forty. In contrast, the participants who did not receive the preschool program were significantly more likely to have been arrested five or more times by age forty.

Not surprisingly, we can't always have the Ferrari. The research equivalent of a Toyota is a cross-sectional data set, which is a collection of data gathered at a single point in time. For example, if epidemiologists are searching for the cause of a new disease (or an outbreak of an old one), they may gather data from all those afflicted in hopes of finding a pattern that leads to the source. What have they eaten? Where have they traveled? What else do they have in common? Researchers may also gather data from individuals who are not afflicted by the disease to highlight contrasts between the two groups.

In fact, all of this exciting cross-sectional data talk reminds me of the week before my wedding, when I became part of a data set. I was working in Kathmandu, Nepal, when I tested positive for a poorly understood stomach illness called "blue-green algae," which had been found in only two places in the world. Researchers had isolated the pathogen that caused the disease, but they were not yet sure what kind of organism it was, as it had never been identified before. When I called home to inform my fiancée about my diagnosis, I acknowledged that there was some bad news. The disease had no known means of transmission, no known cure, and could cause extreme fatigue and other unpleasant side effects for anywhere from a few days to many months.* With the wedding only one week away, yes, this could be a problem. Would I have complete control of my digestive system as I walked down the aisle? Maybe.

But then I really tried to focus on the good news. First, "blue-green algae" was thought to be nonfatal. And second, experts in tropical diseases from as far away as Bangkok had taken a personal interest in my case. *How cool is that?* (Also, I did a terrific job of repeatedly steering the discussion back to the wedding planning: "Enough about my incurable disease. Tell me more about the flowers.")

I spent my final hours in Kathmandu filling out a thirty-page survey describing every aspect of my life: Where did I eat? What did I eat? How did I cook? Did I go swimming? Where and how often? Everyone else who had been diagnosed with the disease was doing the same thing. Eventually the pathogen was identified as a water-borne form of cyanobacteria. (These bacteria are blue, and they are the only kind of bacteria that get their energy from photosynthesis; hence the original description of the disease as "blue-green algae.") The illness was found to respond to treatment with traditional antibiotics, but, curiously, not to some of the newer ones. All of these discoveries were too late to help me, but I was lucky enough to recover quickly anyway. I had near-perfect control of my digestive system by wedding day.

Behind every important study there are good data that made the analysis possible. And behind every bad study . . . well, read on. People often speak about "lying with statistics." I would argue that some of the most egregious statistical mistakes involve *lying with data*; the statistical analysis

* At the time, the disease had a mean duration of forty-three days with a standard deviation of twenty-four days.

is fine, but the data on which the calculations are performed are bogus or inappropriate. Here are some common examples of "garbage in, garbage out."

Selection bias. Pauline Kael, the longtime film critic for *The New Yorker*, is alleged to have said after Richard Nixon's election as president, "Nixon couldn't have won. I don't know anyone who voted for him." The quotation is most likely apocryphal, but it's a lovely example of how a lousy sample (one's group of liberal friends) can offer a misleading snapshot of a larger population (voters from across America). And it introduces the question one should always ask: How have we chosen the sample or samples that we are evaluating? If each member of the relevant population does not have an equal chance of ending up in the sample, we are going to have a problem with whatever results emerge from that sample. One ritual of presidential politics is the Iowa straw poll, in which Republican candidates descend on Ames, Iowa, in August of the year before a presidential election to woo participants, each of whom pays $30 to cast a vote in the poll. The Iowa straw poll does not tell us that much about the future of Republican candidates. (The poll has predicted only three of the last five Republican nominees.) Why? Because Iowans who pay $30 to vote in the straw poll are different from other Iowa Republicans; and Iowa Republicans are different from Republican voters in the rest of the country.

Selection bias can be introduced in many other ways. A survey of consumers in an airport is going to be biased by the fact that people who fly are likely to be wealthier than the general public; a survey at a rest stop on Interstate 90 may have the opposite problem. Both surveys are likely to be biased by the fact that people who are willing to answer a survey in a public place are different from people who would prefer not to be bothered. If you ask 100 people in a public place to complete a short survey, and 60 are willing to answer your questions, *those 60 are likely to be different in significant ways from the 40 who walked by without making eye contact.*

One of the most famous statistical blunders of all time, the notorious *Literary Digest* poll of 1936, was caused by a biased sample. In that year, Kansas governor Alf Landon, a Republican, was running for president against incumbent Franklin Roosevelt, a Democrat. *Literary Digest*, an influential weekly news magazine at the time, mailed a poll to its subscribers and to automobile and telephone owners whose addresses could be culled from public records. All told, the *Literary Digest* poll included 10 million prospective voters, which is an astronomically large sample. As polls with good samples get larger, they get better, since the margin of error shrinks. As polls with bad samples get larger, the pile of garbage just gets bigger and smellier. *Literary Digest* predicted that Landon would beat Roosevelt with 57 percent of the popular vote. In fact, Roosevelt won in a landslide, with 60 percent of the popular vote and forty-six of forty-eight states in the electoral college. The *Literary Digest* sample was "garbage in": the magazine's subscribers were wealthier than average Americans, and therefore more likely to vote Republican, as were households with telephones and cars in 1936.[2]

We can end up with the same basic problem when we compare outcomes between a treatment and a control group if the mechanism for sorting individuals into one group or the other is not random. Consider a recent finding in the medical literature on the side effects of treatment for prostate cancer. There are three common treatments for prostate cancer: surgical removal of the prostate; radiation therapy; or brachytherapy (which involves implanting radioactive "seeds" near the cancer).[3] Impotence is a common side effect of prostate cancer treatment, so researchers have documented the sexual function of men who receive each of the three treatments. A study of 1,000 men found that two years after treatment, 35 percent of the men in the surgery group were able to have sexual intercourse, compared with 37 percent in the radiation group and 43 percent in the brachytherapy group.

Can one look at these data and assume that brachytherapy is least likely to damage a man's sexual function? No, no, no. The authors of the study explicitly warn that we cannot conclude that brachytherapy is better at preserving sexual function, since the men who receive this treatment are generally younger and fitter than men who receive the other treatment. The purpose of the study was merely to document the degree of sexual side effects across all types of treatment.

A related source of bias, known as self-selection bias, will arise whenever individuals volunteer to be in a treatment group. For example, prisoners who volunteer for a drug treatment group are different from other prisoners *because they have volunteered to be in a drug treatment program.* If the participants in this program are more likely to stay out of prison after release than other prisoners, that's great—but it tells us absolutely nothing about the value of the drug treatment program. These former inmates may have changed their lives because the program helped them kick drugs. Or they may have changed their lives because of other factors that also happened to make them more likely to volunteer for a drug treatment program (such as having a really strong desire not to go back to prison). We cannot separate the causal impact of one (the drug treatment program) from the other (being the kind of person who volunteers for a drug treatment program).

Publication bias. Positive findings are more likely to be published than negative findings, which can skew the results that we see. Suppose you have just conducted a rigorous, longitudinal study in which you find conclusively that playing video games *does not* prevent colon cancer. You've followed a representative sample of 100,000 Americans for twenty years; those participants who spend hours playing video games have roughly the same incidence of colon cancer as the participants who do not play video games at all. We'll assume your methodology is impeccable. Which prestigious medical journal is going to publish your results?

None, for two reasons. First, there is no strong scientific reason to believe that playing video games has any impact on colon cancer, so it is not obvious why you were doing this study. Second, and more relevant here, the fact that something *does not* prevent cancer is not a particularly interesting finding. After all, most things *don't* prevent cancer. Negative findings are not especially sexy, in medicine or elsewhere.

The net effect is to distort the research that we see, or do not see. Suppose that one of your graduate school classmates has conducted a different longitudinal study. She finds that people who spend a lot of time playing video games *do* have a lower incidence of colon cancer. *Now that is interesting!* That is exactly the kind of finding that would catch the attention of a medical journal, the popular press, bloggers, and video game makers (who would slap labels on their products extolling the health benefits of their products). It wouldn't be long before Tiger Moms all over the country were "protecting" their children from cancer by snatching books out of their hands and forcing them to play video games instead.

Of course, one important recurring idea in statistics is that unusual things happen every once in a while, just as a matter of chance. If you conduct 100 studies, one of them is likely to turn up results that are pure nonsense—like a statistical association between playing video games and a lower incidence of colon cancer. Here is the problem: The 99 studies that find no link between video games and colon cancer will not get published, because they are not very interesting. The one study that does find a statistical link will make it into print and get loads of follow-on attention. The source of the bias stems not from the studies themselves but from the skewed information that actually reaches the public. Someone reading the scientific literature on video games and cancer would find only a single study, and that single study will suggest that playing video games can prevent cancer. In fact, 99 studies out of 100 would have found no such link.

Yes, my example is absurd—but the problem is real and serious. Here is the first sentence of a *New York Times* article on the publication bias surrounding drugs for treating depression: "The makers of antidepressants like Prozac and Paxil never published the results of about a third of the drug trials that they conducted to win government approval, misleading doctors and consumers about the drugs' true effectiveness."[4] It turns out that 94 percent of studies with positive findings on the effectiveness of these drugs were published, while only 14 percent of the studies with nonpositive results were published. For patients dealing with depression, this is a big deal. When all the studies are included, the antidepressants are better than a placebo by only "a modest margin."

To combat this problem, medical journals now typically require that any study be registered at the beginning of the project if it is to be eligible for publication later on. This gives the editors some evidence on the ratio of positive to nonpositive findings. If 100 studies are registered that propose to examine the effect of skateboarding on heart disease, and only one is ultimately submitted for publication with positive findings, the editors can infer that the other studies had nonpositive findings (or they can at least investigate this possibility).

Recall bias. Memory is a fascinating thing—though not always a great source of good data. We have a natural human impulse to understand the present as a logical consequence of things that happened in the past—cause and effect. The problem is that our memories turn out to be "systematically fragile" when we are trying to explain some particularly good or bad outcome in the present. Consider a study looking at the relationship between diet and cancer. In 1993, a Harvard researcher compiled a data set comprising a group of women with breast cancer and an age-matched group of women who had not been diagnosed with cancer. Women in both groups were asked about their dietary habits earlier in life. The study produced clear results: The women with breast cancer were significantly more likely to have had diets that were high in fat when they were younger.

Ah, but this wasn't actually a study of how diet affects the likelihood of getting cancer. *This was a study of how getting cancer affects a woman's memory of her diet earlier in life.* All of the women in the study had completed a dietary survey years earlier, before any of them had been diagnosed with cancer. The striking finding was that women with breast cancer recalled a diet that was much higher in fat than what they actually consumed; the women with no cancer did not. The *New York Times Magazine* described the insidious nature of this recall bias:

> The diagnosis of breast cancer had not just changed a woman's present and the future; it had altered her past. Women with breast cancer had (unconsciously) decided that a higher-fat diet was a likely predisposition for their disease and (unconsciously) recalled a high-fat diet. It was a pattern poignantly familiar to anyone who knows the history of this stigmatized illness: these women, like thousands of women before them, had searched their own memories for a cause and then summoned that cause into memory.[5]

Recall bias is one reason that longitudinal studies are often preferred to cross-sectional studies. In a longitudinal study the data are collected contemporaneously. At age five, a participant can be asked about his attitudes toward school. Then, thirteen years later, we can revisit that same participant and determine whether he has dropped out of high school. In a cross-sectional study, in which all the data are collected at one point in time, we must ask an eighteen-year-old high school dropout how he or she felt about school at age five, which is inherently less reliable.

Survivorship bias. Suppose a high school principal reports that test scores for a particular cohort of students has risen steadily for four years. The sophomore scores for this class were better than their freshman scores. The scores from junior year were better still, and the senior year scores were best of all. We'll stipulate that there is no cheating going on, and not even any creative use of descriptive statistics. Every year this cohort of students has done better than it did the preceding year, by every possible measure: mean, median, percentage of students at grade level, and so on.

Would you (a) nominate this school leader for "principal of the year" or (b) demand more data?

I say "b." I smell survivorship bias, which occurs when some or many of the observations are falling out of the sample, changing the composition of the observations that are left and therefore affecting the results of any analysis. Let's suppose that our principal is truly awful. The students in his school are learning nothing; each year half of them drop out. Well, that could do very nice things for the school's test scores—without any individual student testing better. If we make the reasonable assumption that the worst students (with the lowest test scores) are the most likely to drop out, then the average test scores of those students left behind will go up steadily as more and more students drop out. (If you have a room of people with varying heights, forcing the short people to leave will raise the average height in the room, but it doesn't make anyone taller.)

The mutual fund industry has aggressively (and insidiously) seized on survivorship bias to make its returns look better to investors than they really are. Mutual funds typically gauge their performance against a key benchmark for stocks, the Standard & Poor's 500, which is an index of 500 leading public companies in America.* If the S&P 500 is up 5.3 percent for the year, a mutual fund is said to beat the index if it performs better than that, or trail the index if it does worse. One cheap and easy option for investors who don't want to pay a mutual fund manager is to buy an S&P 500 Index Fund, which is a mutual fund that simply buys shares in all 500 stocks in the index. Mutual fund managers like to believe that they are savvy investors, capable of using their knowledge to pick stocks that will perform better than a simple index fund. In fact, it turns out to be relatively hard to beat the S&P 500 for any consistent stretch of time. (The S&P 500 is essentially an average of all large stocks being traded, so just as a matter of math we would expect roughly half the actively managed mutual funds to outperform the S&P 500 in a given year and half to underperform.) Of course, it doesn't look very good to lose to a mindless index that simply buys 500 stocks and holds them. No analysis. No fancy macro forecasting. And, much to the delight of investors, no high management fees.

What is a traditional mutual fund company to do? Bogus data to the rescue! Here is how they can "beat the market" without beating the market. A large mutual company will open many new actively managed funds (meaning that experts are picking the stocks, often with a particular focus or strategy). For the sake of example, let's assume that a mutual fund company opens twenty new funds, each of which has roughly a 50 percent chance of beating the S&P 500 in a given year. (This assumption is consistent with long-term data.) Now, basic probability suggests that only ten of the firm's new funds will beat the S&P 500 the first year; five funds will beat it two years in a row; and two or three will beat it three years in a row.

Here comes the clever part. At that point, the new mutual funds with unimpressive returns relative to the S&P 500 are quietly closed.

* The S&P 500 is a nice example of what an index can and should do. The index is made up of the share prices of the 500 leading U.S. companies, each weighted by its market value (so that bigger companies have more weight in the index than smaller companies). The index is a simple and accurate gauge of what is happening to the share prices of the largest American companies at any given time.

(Their assets are folded into other existing funds.) The company can then heavily advertise the two or three new funds that have "consistently outperformed the S&P 500"—even if that performance is the stock-picking equivalent of flipping three heads in a row. The subsequent performance of these funds is likely to revert to the mean, albeit after investors have piled in. The number of mutual funds or investment gurus who have consistently beaten the S&P 500 over a long period is shockingly small.*

Healthy user bias. People who take vitamins regularly are likely to be healthy—*because they are the kind of people who take vitamins regularly!* Whether the vitamins have any impact is a separate issue. Consider the following thought experiment. Suppose public health officials promulgate a theory that all new parents should put their children to bed only in purple pajamas, because that helps stimulate brain development. Twenty years later, longitudinal research confirms that having worn purple pajamas as a child does have an overwhelmingly large positive association with success in life. We find, for example, that 98 percent of entering Harvard freshmen wore purple pajamas as children (and many still do) compared with only 3 percent of inmates in the Massachusetts state prison system.

Of course, the purple pajamas do not matter; but having the kind of parents who put their children in purple pajamas *does matter*. Even when we try to control for factors like parental education, we are still going to be left with unobservable differences between those parents who obsess about putting their children in purple pajamas and those who don't. As *New York Times* health writer Gary Taubes explains, "At its simplest, the problem is that people who faithfully engage in activities that are good for them—taking a drug as prescribed, for instance, or eating what they believe is a healthy diet—are fundamentally different from those who don't."[6] This effect can potentially confound any study trying to evaluate the real effect of activities perceived to be healthful, such as exercising

* For a very nice discussion of why you should probably buy index funds rather than trying to beat the market, read *A Random Walk Down Wall Street*, by my former professor Burton Malkiel.

regularly or eating kale. We think we are comparing the health effects of two diets: kale versus no kale. In fact, if the treatment and control groups are not randomly assigned, we are comparing two diets that are being eaten by two different kinds of people. We have a treatment group that is different from the control group in two respects, rather than just one.

If statistics is detective work, then the data are the clues. My wife spent a year teaching high school students in rural New Hampshire. One of her students was arrested for breaking into a hardware store and stealing some tools. The police were able to crack the case because (1) it had just snowed and there were tracks in the snow leading from the hardware store to the student's home; and (2) the stolen tools were found inside. Good clues help.

Like good data. But first you have to get good data, and that is a lot harder than it seems.

CHAPTER 8

The Central Limit Theorem

The Lebron James of statistics

At times, statistics seems almost like magic. We are able to draw sweeping and powerful conclusions from relatively little data. Somehow we can gain meaningful insight into a presidential election by calling a mere one thousand American voters. We can test a hundred chicken breasts for salmonella at a poultry processing plant and conclude from that sample alone that the entire plant is safe or unsafe. *Where does this extraordinary power to generalize come from?*

Much of it comes from the central limit theorem, which is the Lebron James of statistics—if Lebron were also a supermodel, a Harvard professor, and the winner of the Nobel Peace Prize. The central limit theorem is the "power source" for many of the statistical activities that involve using a sample to make inferences about a large population (like a poll, or a test for salmonella). These kinds of inferences may seem mystical; in fact, they are just a combination of two tools that we've already explored: probability and proper sampling. Before plunging into the mechanics of the central limit theorem (which aren't all that tricky), here is an example to give you the general intuition.

Suppose you live in a city that is hosting a marathon. Runners from all over the world will be competing, which means that many of them do not speak English. The logistics of the race require that runners check in on the morning of the race, after which they are randomly assigned to buses to take them to the starting line. Unfortunately one of the buses gets lost on the way to the race. (Okay, you're going to have to assume that no one has a cell phone and that the driver does not have a GPS navigation device; unless you want to do a lot of unpleasant math right now, just go with it.) As a civic leader in this city, you join the search team.

As luck would have it, you stumble upon a broken-down bus near your home with a large group of unhappy international passengers, none of whom speaks English. This must be the missing bus! You're going to be a hero! Except you have one lingering doubt . . . the passengers on this bus are, well, very large. Based on a quick glance, you reckon that the average weight for this group of passengers has got to be over 220 pounds. There is no way that a random group of marathon runners could all be this heavy. You radio your message to search headquarters: "I think it's the wrong bus. Keep looking."

Further analysis confirms your initial impression. When a translator arrives, you discover that this disabled bus was headed to the International Festival of Sausage, which is also being hosted by your city on the same weekend. (For the sake of verisimilitude, it is entirely possible that sausage festival participants might also be wearing sweat pants.)

Congratulations. If you can grasp how someone who takes a quick look at the weights of passengers on a bus can infer that they are probably not on their way to the starting line of a marathon, then you now understand the basic idea of the central limit theorem. The rest is just fleshing out the details. And if you understand the central limit theorem, most forms of statistical inference will seem relatively intuitive.

The core principle underlying the central limit theorem is that a large, properly drawn sample will resemble the population from which it is drawn. Obviously there will be variation from sample to sample (e.g., each bus headed to the start of the marathon will have a slightly different mix of passengers), but the probability that any sample will deviate massively from the underlying population is very low. This logic is what enabled your snap judgment when you boarded the broken-down bus and saw the average girth of the passengers on board. Lots of big people run marathons; there are likely to be hundreds of people who weigh over 200 pounds in any given race. But the majority of marathon runners are relatively thin. Thus, the likelihood that so many of the largest runners were randomly assigned to the same bus is very, very low. You could conclude with a reasonable degree of confidence that this was not the missing marathon bus. Yes, you could have been wrong, but probability tells us that most of the time you would have been right.

That's the basic intuition behind the central limit theorem. When we add some statistical bells and whistles, we can quantify the likelihood that you will be right or wrong. For example, we might calculate that in a marathon field of 10,000 runners with a mean weight of 155 pounds, there is less than a 1 in 100 chance that a random sample of 60 of those runners (our lost bus) would have a mean weight of 220 pounds or more. For now, let's stick with the intuition; there will be plenty of time for calculations later. The central limit theorem enables us to make the following inferences, all of which will be explored in greater depth in the next chapter.

1. If we have detailed information about some population, then we can make powerful inferences about any properly drawn sample from that population. For example, assume that a school principal has detailed information on the standardized test scores for all the students in his school (mean, standard deviation, etc.). That is the relevant population. Now assume that a bureaucrat from the school district will be arriving next week to give a similar standardized test to 100 randomly selected students. The performance of those 100 students, the sample, will be used to evaluate the performance of the school overall.

   How much confidence can the principal have that the performance of those randomly chosen 100 students will accurately reflect how the entire student body has been performing on similar standardized tests? Quite a bit. According to the central limit theorem, the average test score for the random sample of 100 students will not typically deviate sharply from the average test score for the whole school.
2. If we have detailed information about a properly drawn sample (mean and standard deviation), we can make strikingly accurate inferences about the population from which that sample was drawn. This is essentially working in the opposite direction from the example above, putting ourselves in the shoes of the school district bureaucrat who is evaluating various schools in the dis-

trict. Unlike the school principal, this bureaucrat does not have (or does not trust) the standardized test score data that the principal has for all the students in a particular school, which is the relevant population. Instead, he will be administering a similar test of his own to a random sample of 100 students in each school.

Can this administrator be reasonably certain that the overall performance of any given school can be evaluated fairly based on the test scores of a sample of just 100 students from that school? Yes. The central limit theorem tells us that a large sample will not typically deviate sharply from its underlying population—which means that the sample results (scores for the 100 randomly chosen students) are a good proxy for the results of the population overall (the student body at a particular school). Of course, this is how polling works. A methodologically sound poll of 1,200 Americans can tell us a great deal about how the entire country is thinking.

Think about it: if no. 1 above is true, no. 2 must also be true—and vice versa. If a sample usually looks like the population from which it's drawn, it must also be true that a population will usually look like a sample drawn from that population. (If children typically look like their parents, parents must also typically look like their children.)

3. If we have data describing a particular sample, and data on a particular population, we can infer whether or not that sample is consistent with a sample that is likely to be drawn from that population. This is the missing-bus example described at the beginning of the chapter. We know the mean weight (more or less) for the participants in the marathon. And we know the mean weight (more or less) for the passengers on the broken-down bus. The central limit theorem enables us to calculate the probability that a particular sample (the rotund people on the bus) was drawn from a given population (the marathon field). If that probability is low, then we can conclude with a high degree of confidence that the sample was not drawn from the population in question (e.g., the people on this bus really don't look like a group of marathon runners headed to the starting line).
4. Last, if we know the underlying characteristics of two samples,

we can infer whether or not both samples were likely drawn from the same population. Let us return to our (increasingly absurd) bus example. We now know that a marathon is going on in the city, as well as the International Festival of Sausage. Assume that both groups have thousands of participants, and that both groups are operating buses, all loaded with random samples of either marathon runners or sausage enthusiasts. Further assume that two buses collide. (I already conceded that the example is absurd, so just read on.) In your capacity as a civic leader, you arrive on the scene and are tasked with determining whether or not both buses were headed to the same event (sausage festival or marathon). Miraculously, no one on either bus speaks English, but paramedics provide you with detailed information on the weights of all the passengers on each bus.

From that alone, you can infer whether the two buses were likely headed to the same event, or to different events. Again, think about this intuitively. Suppose that the average weight of the passengers on one bus is 157 pounds, with a standard deviation of 11 pounds (meaning that a high proportion of the passengers weigh between 146 pounds and 168 pounds). Now suppose that the passengers on the second bus have a mean weight of 211 pounds with a standard deviation of 21 pounds (meaning that a high proportion of the passengers weigh between 190 pounds and 232 pounds). Forget statistical formulas for a moment, and just use logic: Does it seem likely that the passengers on those two buses were randomly drawn from the same population?

No. It seems far more likely that one bus is full of marathon runners and the other bus is full of sausage enthusiasts. In addition to the difference in average weight between the two buses, you can also see that the variation in weights *between* the two buses is very large compared with the variation in weights *within* each bus. The folks who weigh one standard deviation above the mean on the "skinny" bus are 168 pounds, which is less than the folks who are one standard deviation below the mean on the "other" bus (190 pounds). This is a telltale sign (both statistically and logically) that the two samples likely came from different populations.

If all of this makes intuitive sense, then you are 93.2 percent of the way to understanding the central limit theorem.* We need to go one step further to put some technical heft behind the intuition. Obviously when you stuck your head inside the broken-down bus and saw a group of large people in sweatpants, you had a "hunch" that they weren't marathoners. The central limit theorem allows us to go beyond that hunch and assign a degree of confidence to your conclusion.

For example, some basic calculations will enable me to conclude that 99 times out of 100 the mean weight of any randomly selected bus of marathoners will be within nine pounds of the mean weight of the entire marathon field. That's what gives statistical heft to my hunch when I stumble across the broken-down bus. These passengers have a mean weight that is twenty-one pounds higher than the mean weight for the marathon field, something that should only occur by chance less than 1 time in 100. As a result, I can reject the hypothesis that this is a missing marathon bus with 99 percent confidence—meaning I should expect my inference to be correct 99 times out of 100.

And yes, probability suggests that on average I'll be *wrong* 1 time in 100.

This kind of analysis all stems from the central limit theorem, which, from a statistical standpoint, has Lebron James–like power and elegance. According to the central limit theorem, the sample means for any population will be distributed roughly as a normal distribution around the population mean. Hang on for a moment as we unpack that statement.

1. Suppose we have a population, like our marathon field, and we are interested in the weights of its members. Any sample of runners, such as each bus of sixty runners, will have a mean.
2. If we take repeated samples, such as picking random groups of sixty runners from the field over and over, then each of those samples will have its own mean weight. These are the sample means.
3. Most of the sample means will be very close to the population mean. Some will be a little higher. Some will be a little lower. Just

* Note the clever use of false precision here.

as a matter of chance, a very few will be significantly higher than the population mean, and a very few will be significantly lower.

Cue the music, because this is where everything comes together in a powerful crescendo . . .

4. The central limit theorem tells us that the sample means will be distributed roughly as a normal distribution around the population mean. The normal distribution, as you may remember from Chapter 2, is the bell-shaped distribution (e.g., adult men's heights) in which 68 percent of the observations lie within one standard deviation of the mean, 95 percent lie within two standard deviations, and so on.
5. All of this will be true no matter what the distribution of the underlying population looks like. The population from which the samples are being drawn does not have to have a normal distribution in order for the sample means to be distributed normally.

Let's think about some real data, say, the household income distribution in the United States. Household income is not distributed normally in America; instead, it tends to be skewed to the right. No household can earn less than $0 in a given year, so that must be the lower bound for the distribution. Meanwhile, a small group of households can earn staggeringly large annual incomes—hundreds of millions or even billions of dollars in some cases. As a result, we would expect the distribution of household incomes to have a long right tail—something like this:

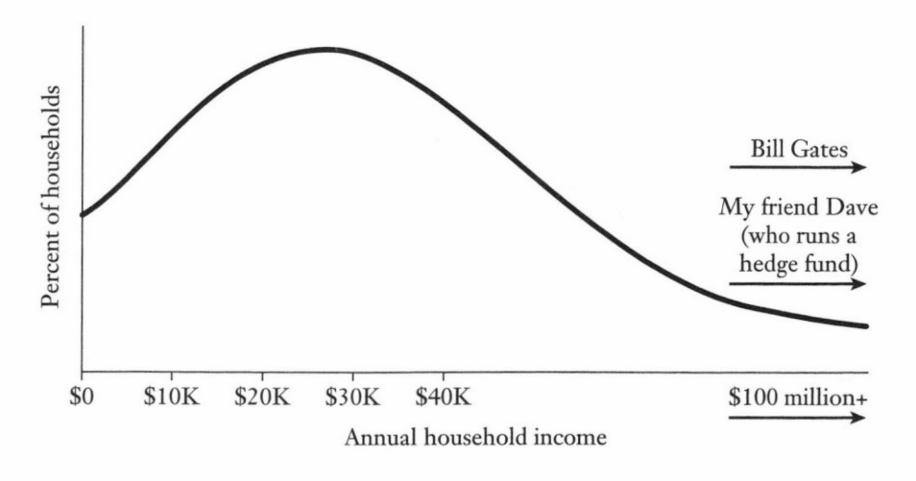

The median household income in the United States is roughly $51,900; the mean household income is $70,900.[1] (People like Bill Gates pull the mean household income to the right, just as he did when he walked in to the bar in Chapter 2.) Now suppose we take a random sample of 1,000 U.S. households and gather information on annual household income. On the basis of the information above, and the central limit theorem, what can we infer about this sample?

Quite a lot, it turns out. First of all, our best guess for what the mean of any sample will be is the mean of the population from which it's drawn. The whole point of a representative sample is that it looks like the underlying population. A properly drawn sample will, on average, look like America. There will be hedge fund managers and homeless people and police officers and everyone else—all roughly in proportion to their frequency in the population. Therefore, we would expect the mean household income for a representative sample of 1,000 American households to be about $70,900. Will it be exactly that? No. But it shouldn't be wildly different either.

If we took multiple samples of 1,000 households, we would expect the different sample means to cluster around the population mean, $70,900. We would expect some means to be higher, and some to be lower. Might we get a sample of 1,000 households with a mean household income of $427,000? Sure, that's possible—but highly unlikely. (Remember, our sampling methodology is sound; we are not conducting a survey in the parking lot of the Greenwich Country Club.) It's also highly unlikely that a proper sample of 1,000 American households would have a mean income of $8,000.

That's all just basic logic. The central limit theorem enables us to go one step further by describing the expected distribution of those different sample means as they cluster around the population mean. Specifically, the sample means will form a normal distribution around the population mean, which in this case is $70,900. Remember, the shape of the underlying population doesn't matter. The household income distribution in the United States is plenty skewed, *but the distribution of the sample means will not be skewed*. If we were to take 100 different samples, each with 1,000 households, and plotted the frequency of our results, we would expect those sample means to form the familiar "bell-shaped" distribution around $70,900.

The larger the number of samples, the more closely the distribution will approximate the normal distribution. And the larger the size of each sample, the tighter that distribution will be. To test this result, let's do a fun experiment with real data on the weights of real Americans. The University of Michigan conducts a longitudinal study called Americans' Changing Lives, which consists of detailed observations on several thousand American adults, including their weights. The weight distribution is skewed slightly right, because it's biologically easier to be 100 pounds overweight than it is to be 100 pounds underweight. The mean weight for all adults in the study is 162 pounds.

Using basic statistical software, we can direct the computer to take a random sample of 100 individuals from the Changing Lives data. In fact, we can do this over and over again to see how the results fit with what the central limit theorem would predict. Here is a graph of the distribution of 100 sample means (rounded to the nearest pound) randomly generated from the Changing Lives data.

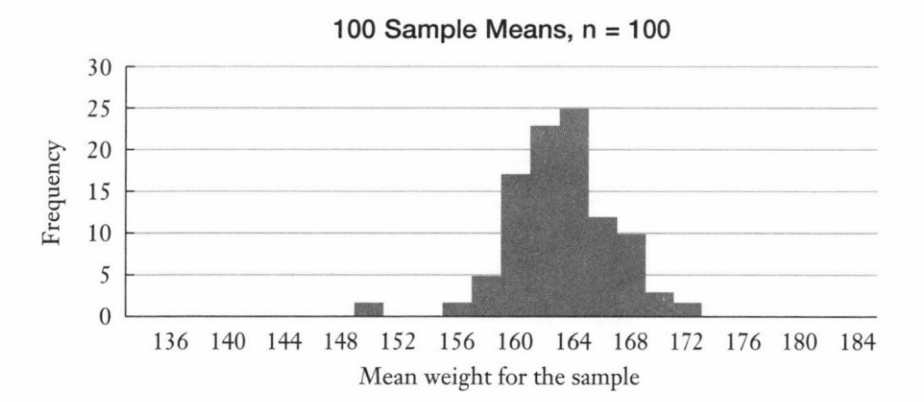

The larger the sample size and the more samples taken, the more closely the distribution of sample means will approximate the normal curve. (As a rule of thumb, the sample size must be at least 30 for the central limit theorem to hold true.) This makes sense. A larger sample is less likely to be affected by random variation. A sample of 2 can be highly skewed by 1 particularly large or small person. In contrast, a sample of 500 will not be unduly affected by a few particularly large or small people.

We are now very close to making all of our statistical dreams come true! The sample means are distributed roughly as a normal curve, as described above. The power of a normal distribution derives from the fact that we know roughly what proportion of observations will lie within one standard deviation above or below the mean (68 percent); what proportion of observations will lie within two standard deviations above or below the mean (95 percent); and so on. This is powerful stuff.

Earlier in this chapter, I pointed out that we could infer intuitively that a busload of passengers with a mean weight twenty-five pounds higher than the mean weight for the whole marathon field was probably not the lost bus of runners. To quantify that intuition—to be able to say that this inference will be correct 95 percent of the time, or 99 percent, or 99.9 percent—we need just one more technical concept: the standard error.

The standard error measures the dispersion of the sample means. How tightly do we expect the sample means to cluster around the population mean? There is some potential confusion here, as we have now introduced two different measures of dispersion: the standard deviation and the standard error. Here is what you need to remember to keep them straight:

1. The standard deviation measures dispersion in the underlying population. In this case, it might measure the dispersion of the weights of all the participants in the Framingham Heart Study, or the dispersion around the mean for the entire marathon field.
2. The standard error measures the dispersion of *the sample means*. If we draw repeated samples of 100 participants from the Framingham Heart Study, what will the dispersion of those sample means look like?
3. Here is what ties the two concepts together: The standard error is the standard deviation of the sample means! Isn't that kind of cool?

A large standard error means that the sample means are spread out widely around the population mean; a small standard error means that they are clustered relatively tightly. Here are three real examples from the Changing Lives data.

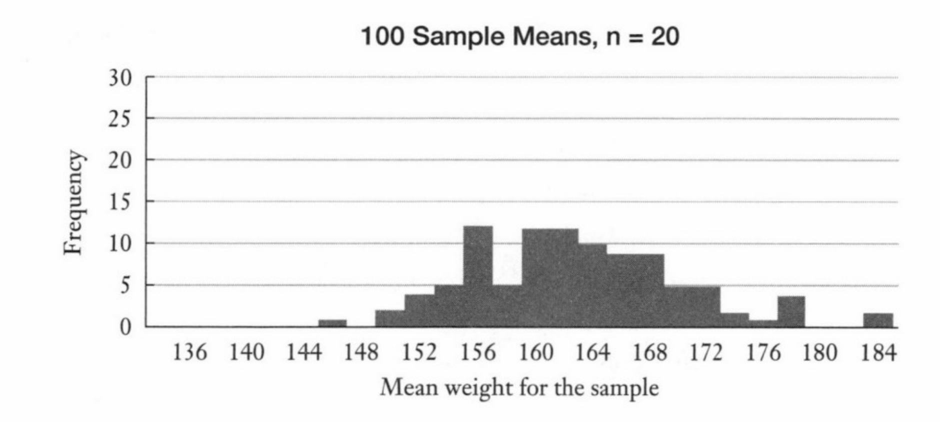

100 Sample Means, n = 100

Frequency

30 25 20 15 10 5 0

136 140 144 148 152 156 160 164 168 172 176 180 184

Mean weight for the sample

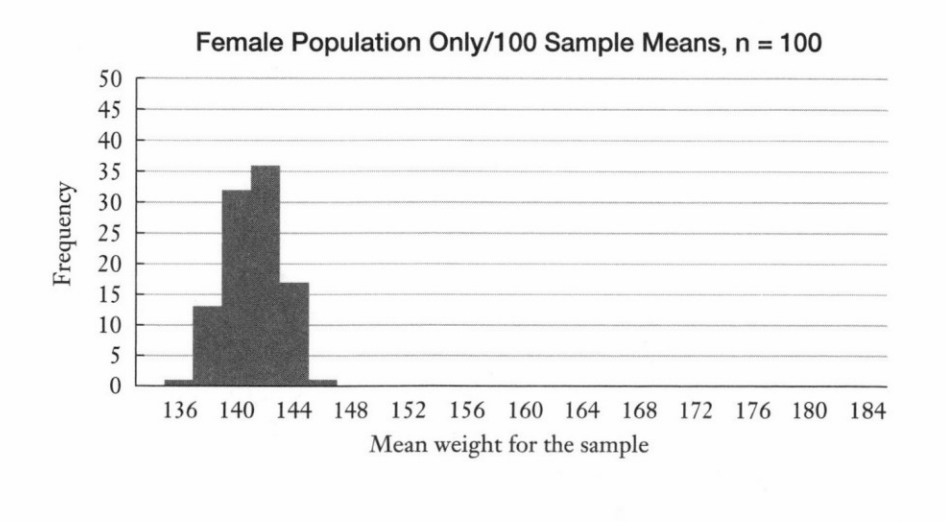

The second distribution, which has a larger sample size, is more tightly clustered around the mean than the first distribution. The larger sample size makes it less likely that a sample mean will deviate sharply from the population mean. The final set of sample means is drawn only from a subset of the population, women in the study. Since the weights of women in the data set are less diffuse than the weights of all persons in the population, it stands to reason that the weights of samples drawn just from the women would be less dispersed than samples drawn from the whole Changing Lives population. (These samples are also clustered around a slightly different population mean, since the mean weight for all females in the Changing Lives study is different from the mean weight for the entire population in the study.)

The pattern that you saw above holds true in general. Sample means will cluster more tightly around the population mean as the size of each sample gets larger (e.g., our sample means were more tightly clustered when we took samples of 100 rather than 30). And the sample means will cluster *less tightly* around the population mean when the underlying population is more spread out (e.g., our sample means for the entire Changing Lives population were more dispersed than the sample means for just the females in the study).

If you've followed the logic this far, then the formula for the standard error follows naturally:

$SE = s/\sqrt{n}$, where s is the standard deviation of the population from which the sample is drawn, and n is the size of the sample. Keep your head about you! Don't let the appearance of letters mess up the basic intuition. The standard error will be large when the standard deviation of the underlying distribution is large. A large sample drawn from a highly dispersed population is also likely to be highly dispersed; a large sample from a population clustered tightly around the mean is also likely to be clustered tightly around the mean. If we are still looking at weight, we would expect the standard error for a sample drawn from the entire Changing Lives population to be larger than the standard error for a sample drawn only from the men in their twenties. *This is why the standard deviation (s) is in the numerator.*

Similarly, we would expect the standard error to get smaller as the sample size gets larger, since large samples are less prone to distortion by extreme outliers. This is why the sample size (n) is in the denominator. (The reason we take the square root of n will be left for a more advanced text; the basic relationship is what's important here.)

In the case of the Changing Lives data, we actually know the standard deviation of the population; often that is not the case. For large samples, we can assume that the standard deviation of the sample is reasonably close to the standard deviation of the population.*

Finally, we have arrived at the payoff for all of this. Because the sample means are distributed normally (thanks to the central limit theorem), we can harness the power of the normal curve. We expect that roughly 68 percent of all sample means will lie within one standard error of the population mean; 95 percent of the sample means will lie within two standard errors of the population mean; and 99.7 percent of the sample means will lie within three standard errors of the population mean.

Frequency Distribution of Sample Means

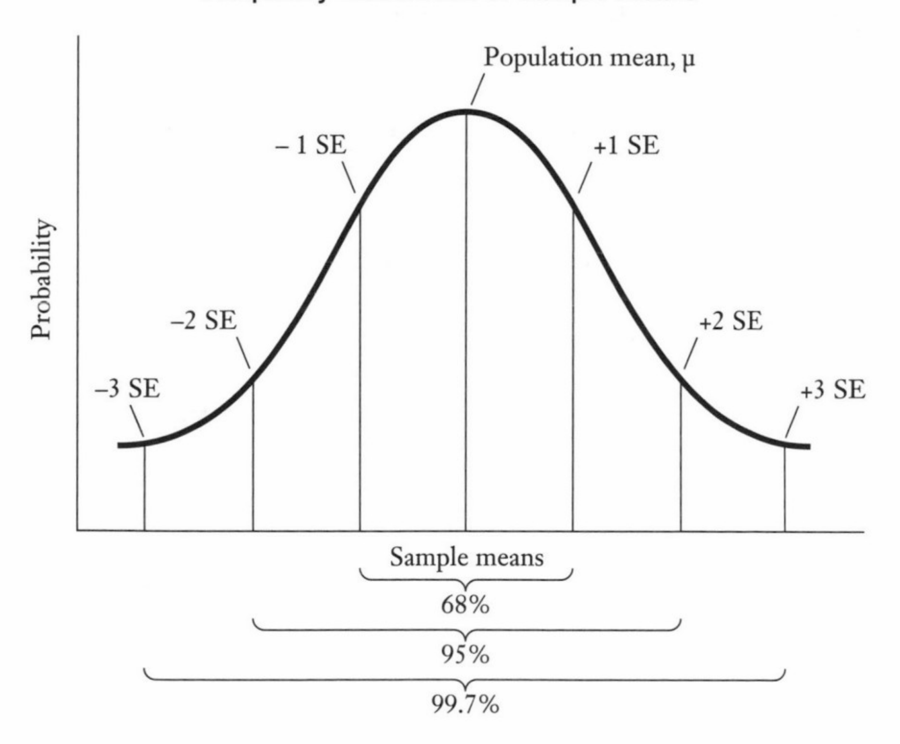

* When the standard deviation for the population is calculated from a smaller sample, the formula is tweaked slightly: $SE = s/\sqrt{n-1}$. This helps to account for the fact that the dispersion in a small sample may understate the dispersion of the full population. This is not highly relevant to the bigger points in this chapter.

So let's return to a variation on our lost-bus example, only now we can substitute numbers for intuition. (The example itself will remain absurd; the next chapter will have plenty of less absurd, real-world examples.) Suppose that the Changing Lives study has invited all of the individuals in the study to meet in Boston for a weekend of data gathering and revelry. The participants are loaded randomly onto buses and ferried among the buildings at the testing facility where they are weighed, measured, poked, prodded, and so on. Shockingly, one bus goes missing, a fact that is broadcast on the local news. At around that time, you are driving back from the Festival of Sausage when you see a crashed bus on the side of the road. Apparently the bus swerved to miss a wild fox crossing the road, and all of the passengers are unconscious but not seriously hurt. (I need them to be uncommunicative for the example to work, but I don't want their injuries to be too disturbing.) Paramedics on the scene inform you that the mean weight of the 62 passengers on the bus is 194 pounds. Also, the fox that the bus swerved to avoid was clipped slightly and appears to have a broken hind leg.

Fortunately you know the mean weight and standard deviation for the entire Changing Lives population, you have a working knowledge of the central limit theorem, *and* you know how to administer first aid to a wild fox. The mean weight for the Changing Lives participants is 162; the standard deviation is 36. From that information, we can calculate the standard error for a 62-person sample (the number of unconscious passengers on the bus): $s/\sqrt{62} = 36/7.9$, or 4.6.

The difference between the sample mean (194 pounds) and the population mean (162 pounds) is 32 pounds, or well more than three standard errors. We know from the central limit theorem that 99.7 percent of all sample means will lie within three standard errors of the population mean. That makes it extremely unlikely that this bus represents a random group of Changing Lives participants. In your duty as a civic leader, you call the study officials to tell them that this is probably not their missing bus, only now you can offer statistical evidence, rather than just "a hunch." You report to the Changing Lives folks that you can reject the possibility that this is the missing bus at the 99.7 percent confidence level. And since you are talking to researchers, they actually understand what you are talking about.

Your analysis is further confirmed when paramedics conduct blood tests on the bus passengers and discover that the mean cholesterol level for the busload of passengers is five standard errors above the mean cholesterol level for the Changing Lives study participants. That suggests, correctly it later turns out, that the unconscious passengers are involved with the Festival of Sausage.

[There is a happy ending. When the bus passengers regained consciousness, Changing Lives study officials offered them counseling on the dangers of a diet high in saturated fats, causing many of them to adopt more heart-healthy eating habits. Meanwhile, the fox was nurtured back to health at a local wildlife preserve and was eventually released back into the wild.]*

I've tried to stick with the basics in this chapter. You should note that for the central limit theorem to apply, the sample sizes need to be relatively large (over 30 as a rule of thumb). We also need a relatively large sample if we are going to assume that the standard deviation of the sample is roughly the same as the standard deviation of the population from which it is drawn. There are plenty of statistical fixes that can be applied when these conditions are not met—but that is all frosting on the cake (and maybe even sprinkles on the frosting on the cake). The "big picture" here is simple and massively powerful:

1. If you draw large, random samples from any population, the means of those samples will be distributed normally around the population mean (regardless of what the distribution of the underlying population looks like).

---

* My University of Chicago colleague Jim Sallee makes a very important critique of the missing-bus examples. He points out that very few buses ever go missing. So if we happen to be looking for a missing bus, any bus that turns up lost or crashed is likely to be that bus, *regardless of the weight of the passengers on the bus*. He's right. (Think about it: if you lose your child in a supermarket, and the store manager tells you that there happens to be a lost child standing near register six, you would conclude immediately that it's probably your child.) We're therefore going to have to add one more element of absurdity to these examples and pretend that buses go missing all the time.

2. Most sample means will lie reasonably close to the population mean; the standard error is what defines "reasonably close."
3. The central limit theorem tells us the probability that a sample mean will lie within a certain distance of the population mean. It is relatively unlikely that a sample mean will lie more than two standard errors from the population mean, and extremely unlikely that it will lie three or more standard errors from the population mean.
4. The less likely it is that an outcome has been observed by chance, the more confident we can be in surmising that some other factor is in play.

That's pretty much what statistical inference is about. The central limit theorem is what makes most of it possible. And until Lebron James wins as many NBA championships as Michael Jordan (six), the central limit theorem will be far more impressive than he is.

CHAPTER 9

Inference

Why my statistics professor thought I might have cheated

In the spring of my senior year of college, I took a statistics class. I wasn't particularly enamored of statistics or of most math-based disciplines at that time, but I had promised my dad that I would take the course if I could leave school for ten days to go on a family trip to the Soviet Union. So, I basically took stats in exchange for a trip to the USSR. This turned out to be a great deal, both because I liked statistics more than I thought I would and because I got to visit the USSR in the spring of 1988. Who knew that the country wouldn't be around in its communist form for much longer?

This story is actually relevant to the chapter; the point is that I wasn't as devoted to my statistics course during the term as I might have been. Among other responsibilities, I was also writing a senior honors thesis that was due about halfway through the term. We had regular quizzes in the statistics course, many of which I ignored or failed. I studied a little for the midterm and did passably well—literally. But a few weeks before the end of the term, two things happened. First, I finished my thesis, giving me copious amounts of new free time. And second, I realized that statistics wasn't nearly as difficult as I had been making it out to be. I began studying the stats book and doing the work from earlier in the course. I earned an A on the final exam.

That's when my statistics professor, whose name I've long since forgotten, called me into his office. I don't remember exactly what he said, but it was something along the lines of "You really did much better on the final than you did on the midterm." This was not a congratulatory visit during which I was recognized for finally doing serious work in the class. There was an implicit accusation (though not an explicit one) in his summons; the expectation was that I would explain why I did so much better on the final exam than the midterm. In short, this guy suspected that I might have cheated. Now that I've taught for many years, I'm more sympathetic to his line of thinking. In nearly every course I've taught, there is a striking degree of correlation between a student's performance on the midterm and on the final. It *is* highly unusual for a student to score below average on the midterm and then near the top of the class on the final.

I explained that I had finished my thesis and gotten serious about the class (by doing things like reading the assigned textbook chapters and doing the homework). He seemed content with this explanation, and I left, still somewhat unsettled by the implicit accusation.

Believe it or not, this anecdote embodies much of what you need to know about statistical inference, including both its strengths and its potential weaknesses. *Statistics cannot prove anything with certainty*. Instead, the power of statistical inference derives from observing some pattern or outcome and then using probability to determine the most likely explanation for that outcome. Suppose a strange gambler arrives in town and offers you a wager: He wins $1,000 if he rolls a six with a single die; you win $500 if he rolls anything else—a pretty good bet from your standpoint. He then proceeds to roll ten sixes in a row, taking $10,000 from you.

One possible explanation is that he was lucky. An alternative explanation is that he cheated somehow. The probability of rolling ten sixes in a row with a fair die is roughly 1 in 60 million. You can't prove that he cheated, but you ought at least to inspect the die.

Of course, the most likely explanation is not always the right explanation. Extremely rare things happen. Linda Cooper is a South Carolina woman who has been struck by lightning four times.[1] (The Federal Emergency Management Administration estimates the probability of getting hit by lightning just once as 1 in 600,000.) Linda Cooper's insurance company cannot deny her coverage simply because her injuries are statistically improbable. To return to my undergraduate statistics exam, the professor had reasonable cause to be suspicious. He saw a pattern that was highly unlikely; this is exactly how investigators spot cheating on standardized exams and how the SEC catches insider trading. But an unlikely pattern is just an unlikely pattern unless it is corroborated by additional evidence. Later in the chapter we will discuss errors that can arise when probability steers us wrong.

For now, we should appreciate that statistical inference uses data to address important questions. Is a new drug effective in treating heart disease? Do cell phones cause brain cancer? Please note that I'm not claiming that statistics can *answer* these kinds of questions unequivocally; instead, inference tells us what is likely, and what is unlikely. Researchers cannot prove that a new drug is effective in treating heart disease, even when they have data from a carefully controlled clinical trial. After all, it is entirely possible that there will be random variation in the outcomes of patients in the treatment and control groups that are unrelated to the new drug. If 53 out of 100 patients taking the new heart disease medication showed marked improvement compared with 49 patients out of 100 receiving a placebo, we would not immediately conclude that the new medication is effective. This is an outcome that can easily be explained by chance variation between the two groups rather than by the new drug.

But suppose instead that 91 out of 100 patients receiving the new drug show marked improvement, compared with 49 out of 100 patients in the control group. It is still possible that this impressive result is unrelated to the new drug; the patients in the treatment group may be particularly lucky or resilient. *But that is now a much less likely explanation*. In the formal language of statistical inference, researchers would likely conclude the following: (1) If the experimental drug has no effect, we would rarely see this amount of variation in outcomes between those who are receiving the drug and those who are taking the placebo. (2) It is therefore highly improbable that the drug has no positive effect. (3) The alternative—and more likely—explanation for the pattern of data observed is that the experimental drug has a positive effect.

Statistical inference is the process by which the data speak to us, enabling us to draw meaningful conclusions. This is the payoff! The point of statistics is not to do myriad rigorous mathematical calculations; the point is to gain insight into meaningful social phenomena. Statistical inference is really just the marriage of two concepts that we've already discussed: data and probability (with a little help from the central limit theorem). I have taken one major methodological shortcut in this chapter. All of the examples will assume that we are working with large, properly drawn samples. This assumption means that the central limit theorem applies, and that the mean and standard deviation for any sample will be roughly the same as the mean and standard deviation for the population from which it is drawn. Both of these things make our calculations easier.

Statistical inference is not dependent on this simplifying assumption, but the assorted methodological fixes for dealing with small samples or imperfect data often get in the way of understanding the big picture. The purpose here is to introduce the power of statistical inference and to explain how it works. Once you get that, it's easy enough to layer on complexity.

One of the most common tools in statistical inference is hypothesis testing. Actually, I've already introduced this concept—just without the fancy terminology. As noted above, statistics alone cannot *prove* anything; instead, we use statistical inference to accept or reject explanations on the basis of their relative likelihood. To be more precise, any statistical inference begins with an implicit or explicit null hypothesis. This is our starting assumption, which will be rejected or not on the basis of subsequent statistical analysis. If we reject the null hypothesis, then we typically accept some alternative hypothesis that is more consistent with the data observed. For example, in a court of law the starting assumption, or null hypothesis, is that the defendant is innocent. The job of the prosecution is to persuade the judge or jury to reject that assumption and accept the alternative hypothesis, which is that the defendant is guilty. As a matter of logic, the alternative hypothesis is a conclusion that must be true if we can reject the null hypothesis. Consider some examples.

Null hypothesis: This new experimental drug is no more effective at preventing malaria than a placebo.

Alternative hypothesis: This new experimental drug can help to prevent malaria.

The data: One group is randomly chosen to receive the new experimental drug, and a control group receives a placebo. At the end of some period of time, the group receiving the experimental drug has far fewer cases of malaria than the control group. This would be an extremely unlikely outcome if the experimental drug had no medical impact. As a result, we *reject* the null hypothesis that the new drug has no impact (beyond that of a placebo), and we accept the logical alternative, which is our alternative hypothesis: This new experimental drug can help to prevent malaria.

This methodological approach is strange enough that we should do one more example. Again, note that the null hypothesis and alternative hypothesis are logical complements. If one is true, the other is not true. Or, if we reject one statement, we must accept the other.

Null hypothesis: Substance abuse treatment for prisoners does not reduce their rearrest rate after leaving prison.

Alternative hypothesis: Substance abuse treatment for prisoners will make them less likely to be rearrested after they are released.

The (hypothetical) data: Prisoners were randomly assigned into two groups; the "treatment" group received substance abuse treatment and the control group did not. (This is one of those cool occasions when the treatment group actually gets treatment!) At the end of five years, both groups have similar rearrest rates. In this case, we *cannot reject* the null hypothesis.* The data have given us no reason to discard our beginning

---

* As a matter of semantics, we have not *proved* the null hypothesis to be true (that substance abuse treatment has no effect). It may turn out to be extremely effective for another group of prisoners. Or perhaps many more of the prisoners in this treatment group would have been rearrested if they had not received the treatment. In any case, on the basis of the data collected, we have merely *failed to reject* our null hypothesis. There is a similar distinction between "failing to reject" a null hypothesis and accepting that null hypothesis. Just because one study could not disprove that substance abuse treatment has no effect (yes, a double negative) does not mean that one must accept that substance abuse treatment is useless. There is a meaningful statistical distinction here. That said, research is often designed to inform policy, and prison officials, who have to decide where to allocate resources, might reasonably accept the position that substance treatment is ineffective until they are persuaded otherwise. Here, as in so many other areas of statistics, judgment matters.

assumption that substance abuse treatment is not an effective tool for keeping ex-offenders from returning to prison.

It may seem counterintuitive, but researchers often create a null hypothesis in hopes of being able to reject it. In both of the examples above, a research "success" (finding a new malaria drug or reducing recidivism) involved rejecting the null hypothesis. The data made that possible in only one of the cases (the malaria drug).

In a courtroom, the threshold for rejecting the presumption of innocence is the qualitative assessment that the defendant is "guilty beyond a reasonable doubt." The judge or jury is left to define what exactly that means. Statistics harnesses the same basic idea, but "guilty beyond a reasonable doubt" is defined quantitatively instead. Researchers typically ask, If the null hypothesis is true, how likely is it that we would observe this pattern of data by chance? To use a familiar example, medical researchers might ask, If this experimental drug has no effect on heart disease (our null hypothesis), how likely is it that 91 out of 100 patients getting the drug would show improvement compared with only 49 out of 100 patients getting a placebo? If the data suggest that the null hypothesis is extremely unlikely—as in this medical example—then we must reject it and accept the alternative hypothesis (that the drug is effective in treating heart disease).

In that vein, let us revisit the Atlanta standardized cheating scandal alluded to at several points in the book. The Atlanta test score results were first flagged because of a high number of "wrong-to-right" erasures. Obviously students taking standardized exams erase answers all the time. And some groups of students may be particularly lucky in their changes, without any cheating necessarily being involved. For that reason, the null hypothesis is that the standardized test scores for any particular school district are legitimate and that any irregular patterns of erasures are merely a product of chance. We certainly do not want to be punishing students or administrators because an unusually high proportion of students happened to make sensible changes to their answer sheets in the final minutes of an important state exam.

But "unusually high" does not begin to describe what was happening in Atlanta. Some classrooms had answer sheets on which the number of wrong-to-right erasures were twenty to fifty standard deviations above the state norm. (To put this in perspective, remember that most observations in a distribution typically fall within two standard deviations of the mean.) So how likely was it that Atlanta students happened to erase massive numbers of wrong answers and replace them with correct answers just as a matter of chance? The official who analyzed the data described the probability of the Atlanta pattern occurring without cheating as roughly equal to the chance of having 70,000 people show up for a football game at the Georgia Dome who all happen to be over seven feet tall.[2] Could it happen? Yes. Is it likely? Not so much.

Georgia officials still could not convict anybody of wrongdoing, just as my professor could not (and should not) have had me thrown out of school because my final exam grade in statistics was out of sync with my midterm grade. *Atlanta officials could not prove that cheating was going on.* They could, however, reject the null hypothesis that the results were legitimate. And they could do so with a "high degree of confidence," meaning that the observed pattern was nearly impossible among normal test takers. They therefore explicitly accepted the alternative hypothesis, which is that something fishy was going on. (I suspect they used more official-sounding language.) Subsequent investigation did in fact uncover the "smoking erasers." There were reports of teachers changing answers, giving out answers, allowing low-scoring children to copy from high-scoring children, and even pointing to answers while standing over students' desks. The most egregious cheating involved a group of teachers who held a weekend pizza party during which they went through exam sheets and changed students' answers.

In the Atlanta example, we could reject the null hypothesis of "no cheating" because the pattern of test results was so wildly improbable in the absence of foul play. But how implausible does the null hypothesis have to be before we can reject it and invite some alternative explanation?

One of the most common thresholds that researchers use for rejecting a null hypothesis is 5 percent, which is often written in decimal form: .05. This probability is known as a significance level, and it represents the upper bound for the likelihood of observing some pattern of data if the null hypothesis were true. Stick with me for a moment, because it's not really that complicated.

Let's think about a significance level of .05. We can reject a null hypothesis at the .05 level if there is less than a 5 percent chance of getting an outcome at least as extreme as what we've observed if the null hypothesis were true. A simple example can make this much clearer. I hate to do this to you, but assume once again that you've been put on missing-bus duty (in part because of your valiant efforts in the last chapter). Only now you are working full-time for the researchers at the Changing Lives study, and they have given you some excellent data to help inform your work. Each bus operated by the organizers of the study has roughly 60 passengers, so we can treat the passengers on any bus as a random sample drawn from the entire Changing Lives population. You are awakened early one morning by the news that a bus in the Boston area has been hijacked by a pro-obesity terrorist group.* Your job is to drop from a helicopter onto the roof of the moving bus, sneak inside through the emergency exit, and then stealthily determine whether the passengers are Changing Lives participants, solely on the basis of their weights. (Seriously, this is no more implausible than most action-adventure plots, and it's a lot more educational.)

As the helicopter takes off from the commando base, you are given a machine gun, several grenades, a watch that also functions as a high-resolution video camera, and the data that we calculated in the last chapter on the mean weight and standard error for samples drawn from the Changing Lives participants. Any random sample of 60 participants will have an expected mean weight of 162 pounds and standard deviation of 36 pounds, since that is the mean and standard deviation for all participants in the study (the population). With those data, we can calculate the standard error for the sample mean: $s/\sqrt{n} = 36/\sqrt{60} = 36/7.75 = 4.6$. At mission control, the following distribution is scanned onto the inside of your right retina, so that you can refer to it after penetrating the moving bus and secretly weighing all the passengers inside.

---

* This example is inspired by real events. Obviously many details have been changed for national security reasons. I can neither confirm nor deny my own involvement.

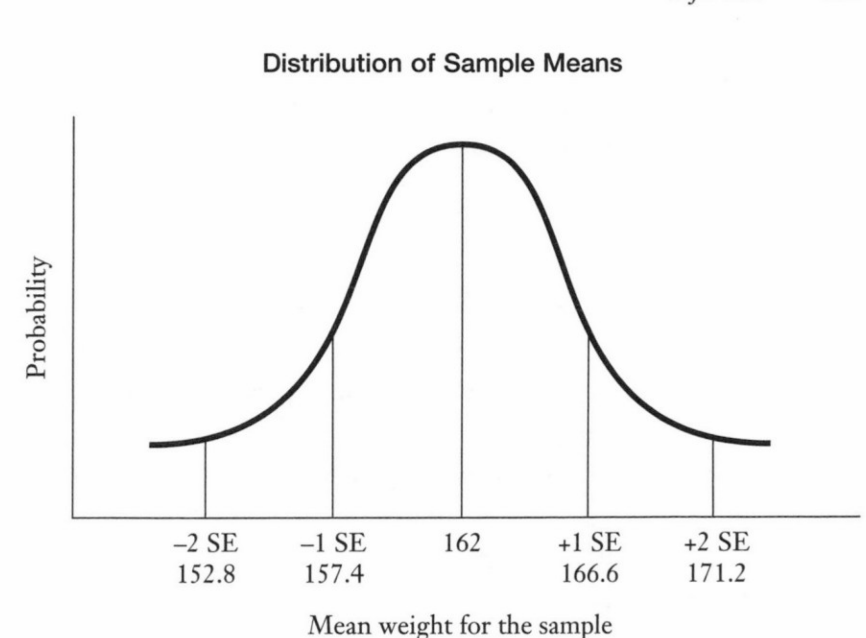

As the distribution above shows, we would expect roughly 95 percent of all 60-person samples drawn from the Changing Lives participants to have a mean weight within two standard errors of the population mean, or roughly between 153 pounds and 171 pounds.* Conversely, only 5 times out of 100 would a sample of 60 persons randomly drawn from the Changing Lives participants have a mean weight that is greater than 171 pounds or less than 153 pounds. (You are conducting what is known as a "two-tailed" hypothesis test; the difference between this and a "one-tailed" test will be covered in an appendix at the end of the chapter.) Your handlers on the counterterrorism task force have decided that .05 is the significance level for your mission. If the mean weight of the 60 passengers on the hijacked bus is above 171 or below 153, then you will reject the null hypothesis that the bus contains Changing Lives participants, accept the alternative hypothesis that the bus contains 60 people headed somewhere else, and await further orders.

---

* To be precise, 95 percent of all sample means will lie within *1.96 standard errors* above or below the population mean.

You successfully drop into the moving bus and secretly weigh all the passengers. The mean weight for this 60-person sample is 136 pounds, which falls more than two standard errors below the mean. (Another important clue is that all of the passengers are children wearing "Glendale Hockey Camp" T-shirts.)

Per your mission instructions, you can reject the null hypothesis that this bus contains a random sample of 60 Changing Lives study participants at the .05 significance level. This means (1) the mean weight on the bus falls into a range that we would expect to observe only 5 times in 100 if the null hypothesis were true and this were really a bus full of Changing Lives passengers; (2) you can reject the null hypothesis at the .05 significance level; and (3) on average, 95 times out of 100 you will have correctly rejected the null hypothesis, and 5 times out of 100 you will be wrong, meaning that you have concluded that this is *not* a bus of Changing Lives participants, when in fact it is. This sample of Changing Lives folks just happens to have a mean weight that is particularly high or low relative to the mean for the study participants overall.

The mission is not quite over. Your handler at mission control (played by Angelina Jolie in the film version of this example) asks you to calculate a p-value for your result. The p-value is the specific probability of getting a result at least as extreme as the one you've observed if the null hypothesis is true. The mean weight for the passengers on this bus is 136, which is 5.7 standard errors below the mean for the Changing Lives study participants. The probability of getting a result at least that extreme if this really were a sample of Changing Lives participants is less than .0001. (In a research document, this would be reported as $p<.0001$.) With your mission complete, you leap from the moving bus and land safely in the passenger seat of a convertible driving in an adjacent lane.

[This story has a happy ending as well. Once the pro–obesity terrorists learn more about your city's International Festival of Sausage, they agree to abandon violence and work peacefully to promote obesity by expanding and promoting sausage festivals around the world.]

• • •

If the .05 significance level seems somewhat arbitrary, that's because it is. There is no single standardized statistical threshold for rejecting a null hypothesis. Both .01 and .1 are also reasonably common thresholds for doing the kind of analysis described above.

Obviously rejecting the null hypothesis at the .01 level (meaning that there is less than a 1 in 100 chance of observing a result in this range if the null hypothesis were true) carries more statistical heft than rejecting the null hypothesis at the .1 level (meaning that there is less than a 1 in 10 chance of observing this result if the null hypothesis were true). The pros and cons of different significance levels will be discussed later in the chapter. For now, the important point is that when we can reject a null hypothesis at some reasonable significance level, the results are said to be "statistically significant."

Here is what that means in real life. When you read in the newspaper that people who eat twenty bran muffins a day have lower rates of colon cancer than people who don't eat prodigious amounts of bran, the underlying academic research probably looked something like this: (1) In some large data set, researchers determined that individuals who ate at least twenty bran muffins a day had a lower incidence of colon cancer than individuals who did not report eating much bran. (2) The researchers' null hypothesis was that eating bran muffins has no impact on colon cancer. (3) The disparity in colon cancer outcomes between those who ate lots of bran muffins and those who didn't could not easily be explained by chance alone. More specifically, if eating bran muffins has no true association with colon cancer, the probability of getting such a wide gap in cancer incidence between bran eaters and non–bran eaters by chance alone is lower than some threshold, such as .05. (This threshold should be established by the researchers *before* they do their statistical analysis to avoid choosing a threshold after the fact that is convenient for making the results look significant.) (4) The academic paper probably contains a conclusion that says something along these lines: "We find a statistically significant association between daily consumption of twenty or more bran muffins and a reduced incidence of colon cancer. These results are significant at the .05 level."

When I subsequently read about that study in the *Chicago Sun-Times* while eating my breakfast of bacon and eggs, the headline is probably more direct and interesting: "20 Bran Muffins a Day Help Keep Colon Cancer Away." However, that newspaper headline, while much more interesting to read than the academic paper, may also be introducing a serious inaccuracy. The study does not actually claim that eating bran muffins lowers an individual's risk of getting colon cancer; it merely shows a negative correlation between the consumption of bran muffins and the incidence of colon cancer in one large data set. This statistical association is not sufficient to prove that the bran muffins *cause* the improved health outcome. After all, the kind of people who eat bran muffins (particularly twenty a day!) may do lots of other things that lower their cancer risk, such as eating less red meat, exercising regularly, getting screened for cancer, and so on. (This is the "healthy user bias" from Chapter 7.) Is it the bran muffins at work here, or is it other behaviors or personal attributes that happen to be shared by people who eat a lot of bran muffins? This distinction between correlation and causation is crucial to the proper interpretation of statistical results. We will revisit the idea that "correlation does not equal causation" later in the book.

I should also point out that statistical significance says nothing about the *size* of the association. People who eat lots of bran muffins may have a lower incidence of colon cancer—but how much lower? The difference in colon cancer rates for bran muffin eaters and non–bran muffin eaters may be trivial; the finding of statistical significance means only that the observed effect, however tiny, is not likely to be a coincidence. Suppose you stumble across a well-designed study that has found a statistically significant positive relationship between eating a banana before the SAT and achieving a higher score on the math portion of the test. One of the first questions you want to ask is, How big is this effect? It could easily be .9 points; on a test with a mean score of 500, that is not a life-changing figure. In Chapter 11, we will return to this crucial distinction between *size* and *significance* when it comes to interpreting statistical results.

Meanwhile, a finding that there is "no statistically significant association" between two variables means that any relationship between the two variables can reasonably be explained by chance alone. The *New York Times* recently ran an exposé on technology companies peddling software that they claim improves student performance, when the data suggest otherwise.[3] According to the article, Carnegie Mellon University sells a software program called Cognitive Tutor with this bold claim: "Revolutionary Math Curricula. Revolutionary Results." Yet an assessment of Cognitive Tutor conducted by the U.S. Department of Education concluded that the product had "no discernible effects" on the test scores of high school students. (The *Times* suggested that the appropriate marketing campaign should be "Undistinguished Math Curricula. Unproven Results.") In fact, a study of ten software products designed to teach skills such as math or reading found that nine of them "did not have statistically significant effects on test scores." In other words, federal researchers cannot rule out mere chance as the cause of any variation in the performance of students who use these software products and students who do not.

Let me pause here to remind you why all of this matters. An article in the *Wall Street Journal* in May of 2011 carried the headline "Link in Autism, Brain Size." This is an important breakthrough, as the causes of autism spectrum disorder remain elusive. The first sentence of the *Wall Street Journal* story, which summarized a paper published in the *Archives of General Psychiatry*, reports, "Children with autism have larger brains than children without the disorder, and the growth appears to occur before age 2, according to a new study released on Monday."[4] On the basis of brain imaging conducted on 59 children with autism and 38 children without autism, researchers at the University of North Carolina reported that children with autism have brains that are up to 10 percent larger than those of children of the same age without autism.

Here is the relevant medical question: Is there a physiological difference in the brains of young children who have autism spectrum disorder? If so, this insight might lead to a better understanding of what causes the disorder and how it can be treated or prevented.

And here is the relevant statistical question: Can researchers make sweeping inferences about autism spectrum disorder in general that are based on a study of a seemingly small group of children with autism (59) and an even smaller control group (38)—a mere 97 subjects in all? The answer is yes. The researchers concluded that the probability of observing the differences in total brain size that they found in their two samples would be a mere 2 in 1,000 (p = .002) if there is in fact no real difference in brain size between children with and without autism spectrum disorder in the overall population.

I tracked down the original study in the *Archives of General Psychiatry*.[5] The methods used by these researchers are no more sophisticated than the concepts we've covered so far. I will give you a quick tour of the underpinnings of this socially and statistically significant result. First, you should recognize that each group of children, the 59 with autism and the 38 without autism, constitutes a reasonably large sample drawn from their respective populations—all children with and without autism spectrum disorder. The samples are large enough that the central limit will apply. If you've already tried to block the last chapter out of your mind, I will remind you of what the central limit theorem tells us: (1) the sample means for any population will be distributed roughly as a normal distribution around the true population mean; (2) we would expect the sample mean and the sample standard deviation to be roughly equal to the mean and standard deviation for the population from which it is drawn; and (3) roughly 68 percent of sample means will lie within one standard error of the population mean, roughly 95 percent will lie within two standard errors of the population mean, and so on.

In less technical language, this all means that any sample should look a lot like the population from which it is drawn; while every sample will be different, it would be relatively rare for the mean of a properly drawn sample to deviate by a huge amount from the mean for the relevant underlying population. Similarly, we would also expect two samples drawn from the same population to look a lot like each other. Or, to think about the situation somewhat differently, if we have two samples that have extremely dissimilar means, the most likely explanation is that they came from different populations.

Here is a quick intuitive example. Suppose your null hypothesis is that male professional basketball players have the same mean height as the rest of the adult male population. You randomly select a sample of 50 professional basketball players and a sample of 50 men who do not play professional basketball. Suppose the mean height of your basketball sample is 6 feet 7 inches, and the mean height of the non–basketball players is 5 feet 10 inches (a 9-inch difference). What is the probability of observing such a large difference in mean height between the two samples if in fact there is no difference in average height between professional basketball players and all other men in the overall population? The nontechnical answer: very, very, very low.*

The autism research paper has the same basic methodology. The paper compares several measures of brain size between the samples of children. (The brain measurements were done with magnetic resonance imaging at age two, and again between ages four and five.) I'll focus on just one measurement, the total brain volume. The researchers' null hypothesis was presumably that there are no anatomical differences in the brains of children with and without autism. The alternative hypothesis is that the brains of children with autism spectrum disorder are fundamentally different. Such a finding would still leave lots of questions, but it would point to a direction for further inquiry.

In this study, the children with autism spectrum disorder had a mean brain volume of 1310.4 cubic centimeters; the children in the control group had a mean brain volume of 1238.8 cubic centimeters. Thus, the difference in average brain volume between the two groups is 71.6 cubic centimeters. How likely would this result be if in fact there were no difference in average brain size in the general population between children who have autism spectrum disorder and children who do not?

You may recall from the last chapter that we can create a standard error for each of our samples: $s/\sqrt{n}$, where s is the standard deviation of the sample and n is the number of observations. The research paper gives us these figures. The standard error for the total brain volume of the 59 children in the autism spectrum disorder sample is 13 cubic centimeters;

---

* There are two possible alternative hypotheses. One is that male professional basketball players are taller than the overall male population. The other is merely that male professional basketball players have a different mean height than the overall male population (leaving open the possibility that male basketball players may actually be shorter than other men). This distinction has a small impact when one performs significance tests and calculates p-values. It is explained in more advanced texts and is not important to our general discussion here.

the standard error for the total brain volume of the 38 children in the control group is 18 cubic centimeters. You will recall that the central limit theorem tells us that for 95 samples out of 100, the sample mean is going to lie within two standard errors of the true population mean, in one direction or the other.

As a result, we can infer from our sample that 95 times out of 100 the interval of 1310.4 cubic centimeters ± 26 (which is two standard errors) will contain the average brain volume for *all* children with autism spectrum disorder. This expression is called a confidence interval. We can say with 95 percent confidence that the range 1284.4 to 1336.4 cubic centimeters contains the average total brain volume for children in the general population with autism spectrum disorder.

Using the same methodology, we can say with 95 percent confidence that the interval of 1238.8 ± 36, or between 1202.8 and 1274.8 cubic centimeters, will include the average brain volume for children in the general population who do not have autism spectrum disorder.

Yes, there are a lot of numbers here. Perhaps you've just hurled the book across the room.* If not, or if you then went and retrieved the book, what you should notice is that our confidence intervals do not overlap. The *lower bound* of our 95 percent confidence interval for the average brain size of children with autism in the general population (1284.4 cubic centimeters) is still higher than the *upper bound* for the 95 percent confidence interval for the average brain size for young children in the population without autism (1274.8 cubic centimeters), as the following diagram illustrates.

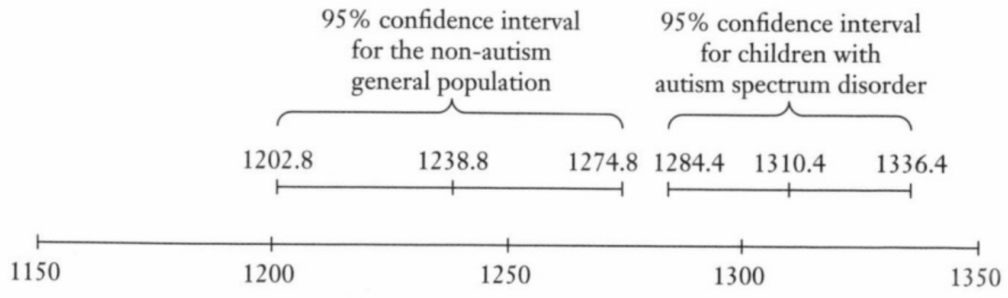

* I will admit that I did once tear a statistics book in half out of frustration.

This is the first clue that there may be an underlying anatomical difference in the brains of young children with autism spectrum disorder. Still, it's just a clue. All of these inferences are based on data from fewer than 100 children. Maybe we just have wacky samples.

One final statistical procedure can bring all this to fruition. If statistics were an Olympic event like figure skating, this would be the last program, after which elated fans throw bouquets of flowers onto the ice. We can calculate the exact probability of observing a difference of means at least this large (1310.4 cubic centimeters versus 1238.8 cubic centimeters) if there is really no difference in brain size between children with autism spectrum and all other children in the general population. We can find a p-value for the observed difference in means.

Lest you hurl the book across the room again, I have put the formula in an appendix. The intuition is quite straightforward. If we draw two large samples from the same population, we would expect them to have very similar means. In fact, our best guess is that they will have identical means. For example, if I were to select 100 players from the NBA and they had an average height of 6 feet 7 inches, then I would expect another random sample of 100 players from the NBA to have a mean height close to 6 feet 7 inches. Okay, maybe the two samples would be an inch or 2 apart. But it's less likely that the means of the two samples will be 4 inches apart—and even less likely that there will be a difference of 6 or 8 inches. It turns out that we can calculate a standard error for the difference between two sample means; this standard error gives us a measure of the dispersion we can expect, on average, when we subtract one sample mean from the other. (As noted earlier, the formula is in the chapter appendix.) The important thing is that we can use this standard error to calculate the probability that two samples come from the same population. Here is how it works:

1. If two samples are drawn from the same population, our best guess for the difference between their means is zero.
2. The central limit theorem tells us that in repeated samples, the *difference between the two means* will be distributed roughly as a normal distribution. (Okay, have you come to love the central limit theorem yet or not?)

3. If the two samples really have come from the same population, then in roughly 68 cases out of 100, the difference between the two sample means will be within one standard error of zero. And in roughly 95 cases out of 100, the difference between the two sample means will be within two standard errors of zero. And in 99.7 cases out of 100, the difference will be within three standard errors of zero—which turns out to be what motivates the conclusion in the autism research paper that we started with.

As noted earlier, the difference in the mean brain size between the sample of children with autism spectrum disorder and the control group is 71.6 cubic centimeters. The standard error on that difference is 22.7, meaning that the difference in means between the two samples is more than three standard errors from zero; we would expect an outcome this extreme (or more so) only 2 times in 1,000 if these samples are drawn from an identical population.

In the paper published in the *Archives of General Psychiatry*, the authors report a p-value of .002, as I mentioned earlier. Now you know where it came from!

For all the wonders of statistical inference, there are some significant pitfalls. They derive from the example that introduced the chapter: my suspicious statistics professor. The powerful process of statistical inference is based on probability, not on some kind of cosmic certainty. We don't want to be sending people to jail just for doing the equivalent of drawing two royal flushes in a row; it *can* happen, even if someone is not cheating. As a result, we have a fundamental dilemma when it comes to any kind of hypothesis testing.

This statistical reality came to a head in 2011 when the *Journal of Personality and Social Psychology* prepared to publish an academic paper that, on the surface, seemed like thousands of other academic papers.[6] A Cornell professor explicitly proposed a null hypothesis, conducted an experiment to test his null hypothesis, and then rejected the null hypothesis at the .05 significance on the basis of the experimental results. The result was uproar, in scientific circles as well as mainstream media outlets like the *New York Times*.

Suffice it to say that articles in the *Journal of Personality and Social Psychology* don't usually attract big headlines. What exactly made this study so controversial? The researcher in question was testing humans' capacity to exercise extrasensory perception, or ESP. The null hypothesis was that ESP does not exist; the alternative hypothesis was that humans do have extrasensory powers. To study this question, the researcher recruited a large sample of participants to examine two "curtains" posted on a computer screen. A software program randomly put an erotic photo behind one curtain or the other. In repeated trials, study participants were able to pick the curtain with the erotic photo behind it 53 percent of the time, whereas probability says they would be right only 50 percent of the time. Because of the large sample size, the researcher was able to reject the null hypothesis that extrasensory perception does not exist and accept instead the alternative hypothesis that extrasensory perception can enable individuals to sense future events. The decision to publish the paper was widely criticized on the grounds that a single statistically significant event can easily be a product of chance, especially when there is no other evidence corroborating or even explaining the finding. The *New York Times* summarized the critiques: "Claims that defy almost every law of science are by definition extraordinary and thus require extraordinary evidence. Neglecting to take this into account—as conventional social science analyses do—makes many findings look far more significant than they really are."

One answer to this kind of nonsense would appear to be a more rigorous threshold for defining statistical significance, such as .001.* But that creates problems of its own. Choosing an appropriate significance level involves an inherent trade-off.

If our burden of proof for rejecting the null hypothesis is too low (e.g., .1), we are going to find ourselves periodically rejecting the null

* Another answer is to attempt to replicate the results in additional studies.

hypothesis when in fact it is true (as I suspect was the case with the ESP study). In statistical parlance, this is known as a Type I error. Consider the example of an American courtroom, where the null hypothesis is that a defendant is not guilty and the threshold for rejecting that null hypothesis is "guilty beyond a reasonable doubt." Suppose we were to relax that threshold to something like "a strong hunch that the guy did it." This is going to ensure that more criminals go to jail—and also more innocent people. In a statistical context, this is the equivalent of having a relatively low significance level, such as .1.

Well, 1 in 10 is not exactly wildly improbable. Consider this challenge in the context of approving a new cancer drug. For every ten drugs that we approve with this relatively low burden of statistical proof, one of them does not actually work and showed promising results in the trial just by chance. (Or, in the courtroom example, for every ten defendants that we find guilty, one of them was actually innocent.) A Type I error involves wrongly rejecting a null hypothesis. Though the terminology is somewhat counterintuitive, this is also known as a "false positive." Here is one way to reconcile the jargon. When you go to the doctor to get tested for some disease, the null hypothesis is that you do not have that disease. If the lab results can be used to reject the null hypothesis, then you are said to test positive. And if you test positive but are not really sick, then it's a false positive.

In any case, the lower our statistical burden for rejecting the null hypothesis, the more likely it is to happen. Obviously we would prefer not to approve ineffective cancer drugs, or send innocent defendants to prison.

But there is a tension here. The higher the threshold for rejecting the null hypothesis, the more likely it is that we will fail to reject a null hypothesis that ought to be rejected. If we require five eyewitnesses in order to convict every criminal defendant, then a lot of guilty defendants are wrongly going to be set free. (Of course, fewer innocents will go to prison.) If we adopt a .001 significance level in the clinical trials for all new cancer drugs, then we will indeed minimize the approval of ineffective drugs. (There is only a 1 in 1,000 chance of wrongly rejecting the null hypothesis that the drug is no more effective than a placebo.) Yet now we introduce the risk of not approving many effective drugs because we have set the bar for approval so high. This is known as a Type II error, or false negative.

Which kind of error is worse? That depends on the circumstances. The most important point is that you recognize the trade-off. There is no statistical "free lunch." Consider these nonstatistical situations, all of which involve a trade-off between Type I and Type II errors.

1. Spam filters. The null hypothesis is that any particular e-mail message is *not* spam. Your spam filter looks for clues that can be used to reject that null hypothesis for any particular e-mail, such as huge distribution lists or phrases like "penis enlargement." A Type I error would involve screening out an e-mail message that is not actually spam (a false positive). A Type II error would involve letting spam through the filter into your inbox (a false negative). Given the costs of missing an important e-mail relative to the costs of getting the occasional message about herbal vitamins, most people would probably err on the side of allowing Type II errors. An optimally designed spam filter should require a relatively high degree of certainty before rejecting the null hypothesis that an incoming e-mail is legitimate and blocking it.
2. Screening for cancer. We have numerous tests for the early detection of cancer, such as mammograms (breast cancer), the PSA test (prostate cancer), and even full-body MRI scans for anything else that might look suspicious. The null hypothesis for anyone undergoing this kind of screening is that no cancer is present. The screening is used to reject this null hypothesis if the results are suspicious. The assumption has always been that a Type I error (a false positive that turns out to be nothing) is far preferable to a Type II error (a false negative that misses a cancer diagnosis). Historically, cancer screening has been the opposite of the spam example. Doctors and patients are willing to tolerate a fair number of Type I errors (false positives) in order to avoid the possibility of a Type II error (missing a cancer diagnosis). More recently, health policy experts have begun to challenge this view because of the high costs and serious side effects associated with false positives.
3. Capturing terrorists. Neither a Type I nor a Type II error is acceptable in this situation, which is why society continues to debate about the appropriate balance between fighting terrorism

> and protecting civil liberties. The null hypothesis is that an individual is not a terrorist. As in a regular criminal context, we do not want to commit a Type I error and send innocent people to Guantánamo Bay. Yet in a world with weapons of mass destruction, letting even a single terrorist go free (a Type II error) can be literally catastrophic. This is why—whether you approve of it or not—the United States is holding suspected terrorists at Guantánamo Bay on the basis of less evidence than might be required to convict them in a regular criminal court.

Statistical inference is not magic, nor is it infallible, but it is an extraordinary tool for making sense of the world. We can gain great insight into many life phenomena just by determining the most likely explanation. Most of us do this all the time (e.g., "I think that college student passed out on the floor surrounded by beer cans has had too much to drink" rather than "I think that college student passed out on the floor surrounded by beer cans has been poisoned by terrorists").

Statistical inference merely formalizes the process.

APPENDIX TO CHAPTER 9

Calculating the standard error for a difference of means

Formula for comparing two means

$$\frac{\bar{x} - \bar{y}}{\sqrt{\frac{s_x^2}{n_x} + \frac{s_y^2}{n_y}}}$$

$\bar{x} - \bar{y}$ ⟶ numerator yields the size of the difference in means

$\sqrt{\frac{s_x^2}{n_x} + \frac{s_y^2}{n_y}}$ ⟶ denominator yields the standard error for a difference in mean between two samples

where $\bar{x}$ = mean for sample x

$\bar{y}$ = mean for sample y

s_x = standard deviation for sample x
s_y = standard deviation for sample y
n_x = number of observations in sample x
n_y = number of observations in sample y

Our null hypothesis is that the two sample means are the same. The formula above calculates the observed difference in means relative to the size of the standard error for the difference in means. Again, we lean heavily on the normal distribution. If the underlying population means are truly the same, then we would expect the difference in sample means to be less than one standard error about 68 percent of the time; less than two standard errors about 95 percent of the time; and so on.

In the autism example from the chapter, the difference in the mean between the two samples was 71.6 cubic centimeters with a standard error of 22.7. The ratio of that observed difference is 3.15, meaning that the two samples have means that are more than 3 standard errors apart. As noted in the chapter, the probability of getting samples with such different means if the underlying populations have the same mean is very, very low. Specifically, the probability of observing a difference of means that is 3.15 standard errors or larger is .002.

Difference in Sample Means

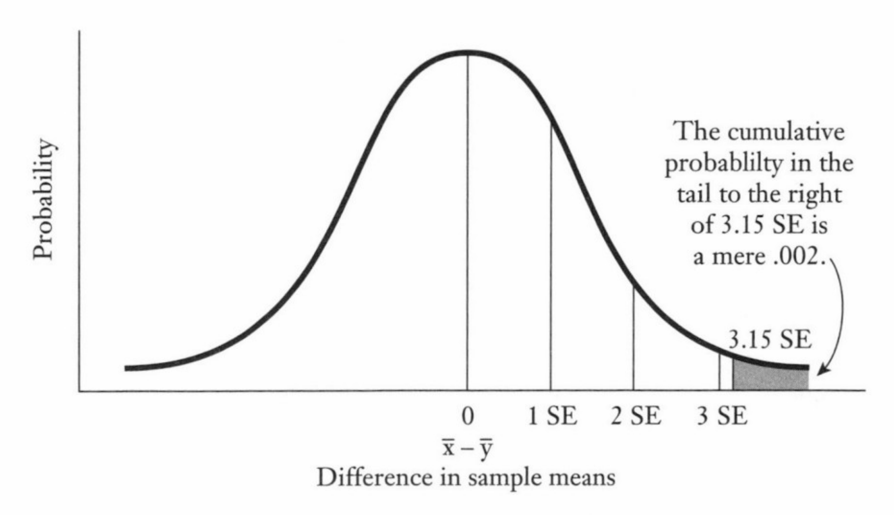

One- and Two-Tailed Hypothesis Testing

This chapter introduced the idea of using samples to test whether male professional basketball players *are the same height* as the general population. I finessed one detail. Our null hypothesis is that male basketball players have the same mean height as men in the general population. What I glossed over is that we have two possible alternative hypotheses.

One alternative hypothesis is that male professional basketball players have a different mean height than the overall male population; they could be taller than other men in the population, or shorter. This was the approach that you took when you dropped into the hijacked bus and weighed the passengers to determine whether they were participants in the Changing Lives study. You could reject the null hypothesis that the bus participants were part of the study if the passengers' mean weight was significantly higher than the overall mean for Changing Lives participants *or* if it was significantly lower (as turned out to be the case). Our second alternative hypothesis is that male professional basketball players are taller on average than other men in the population. In this case, the background knowledge that we bring to this question tells us that basketball players cannot possibly be shorter than the general population. The distinction between these two alternative hypotheses will determine whether we do a one-tailed hypothesis test or a two-tailed hypothesis test.

In both cases, let's assume that we are going to do a significance test at the .05 level. We will reject our null hypothesis if we observe a difference in heights between the two samples that would occur 5 times in 100 or less if all these guys really are the same height. So far, so good.

Here is where things get a little more nuanced. When our alternative hypothesis is that basketball players are taller than other men, we are going to do a *one-tailed hypothesis test*. We will measure the difference in mean height between our sample of male basketball players and our sample of regular men. We know that if our null hypothesis is true, then we will observe a difference that is 1.64 standard errors or greater only 5 times in 100. We reject our null hypothesis if our result falls in this range, as the following diagram shows.

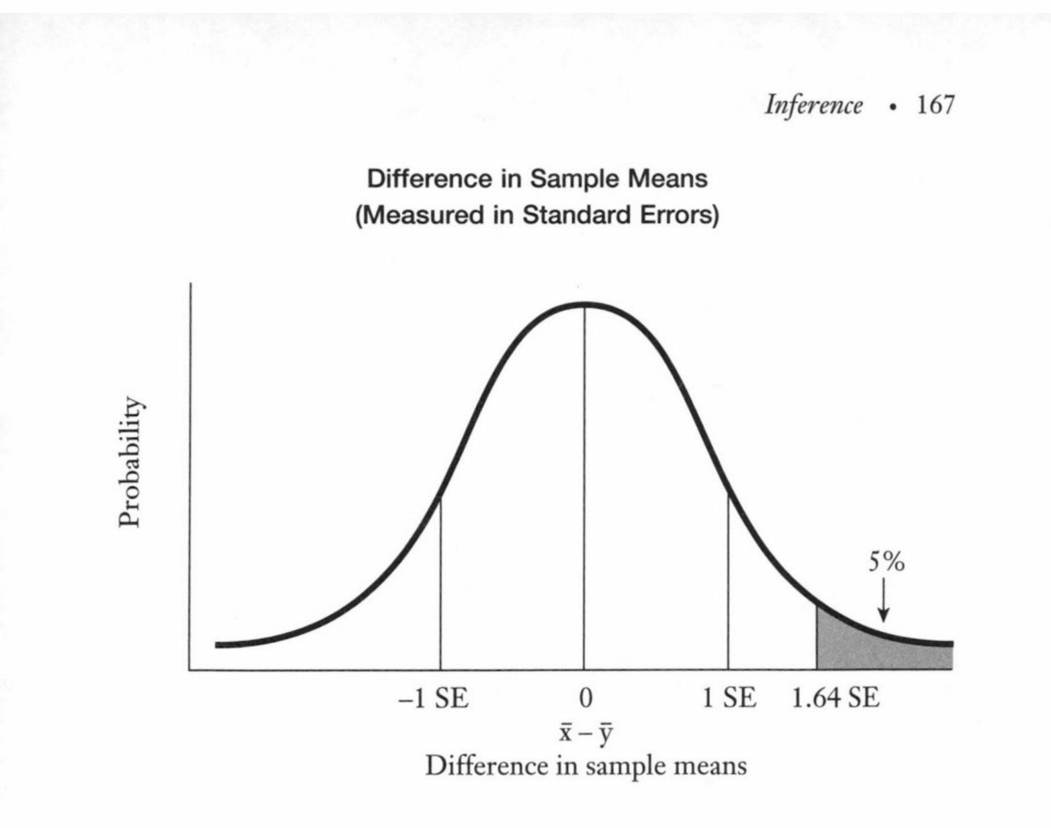

Now let's revisit the other alternative hypothesis—that male basketball players could be taller or shorter than the general population. Our general approach is the same. Again, we will reject our null hypothesis that basketball players are the same height as the general population if we get a result that would occur 5 times in 100 or less if there really is no difference in heights. The difference, however, is that we must now entertain the possibility that basketball players are shorter than the general population. We will therefore reject our null hypothesis if our sample of male basketball players has a mean height that is significantly higher *or lower* than the mean height for our sample of normal men. This requires a *two-tailed hypothesis test*. The cutoff points for rejecting our null hypothesis will be different because we must now account for the possibility of a large difference in sample means in both directions: positive or negative. More specifically, the range in which we will reject our null hypothesis has been split between the two tails. We will still reject our null hypothesis if we get an outcome that would occur 5 percent of the time or less if basketball players are the same height as the general population; only now we have two different ways that we can end up rejecting the null hypothesis.

We will reject our null hypothesis if the mean height for the sample of male basketball players is so much larger than the mean for the normal men that we would observe such an outcome only *2.5 times in 100* if basketball players are really the same height as everyone else.

And we will reject our null hypothesis if the mean height for the sample of male basketball players is so much smaller than the mean for the normal men that we would observe such an outcome only 2.5 times in 100 if basketball players are really the same height as everyone else.

Together, these two contingencies add up to 5 percent, as the diagram below illustrates.

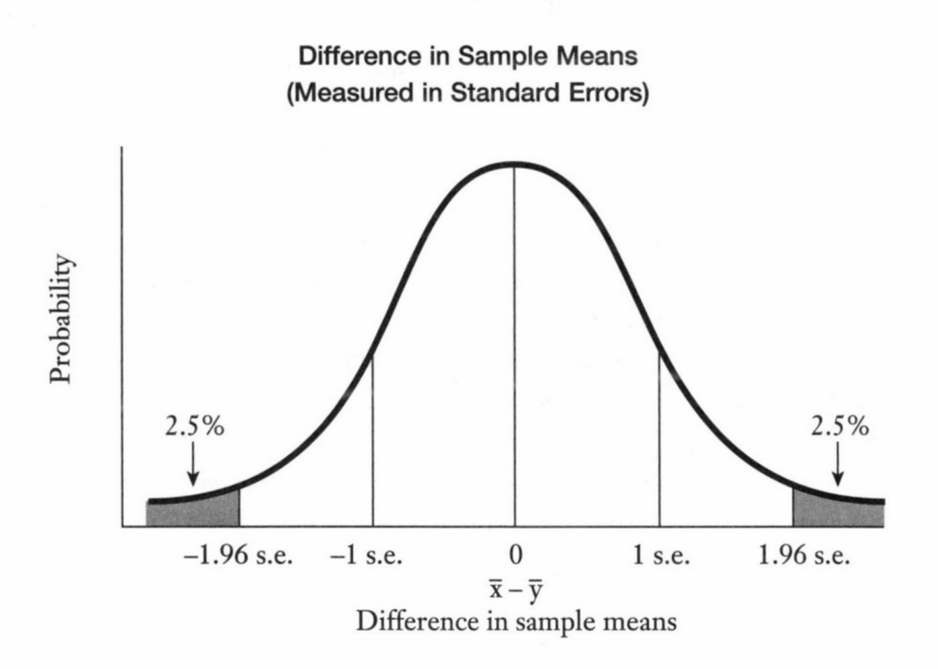

Judgment should inform whether a one- or a two-tailed hypothesis is more appropriate for the analysis being conducted.

CHAPTER 10

Polling

How we know that 64 percent of Americans support the death penalty (with a sampling error ± 3 percent)

In late 2011, the *New York Times* ran a front-page story reporting that "a deep sense of anxiety and doubt about the future hangs over the nation."[1] The story delved into the psyche of America, offering insights into public opinion on topics ranging from the performance of the Obama administration to the distribution of wealth. Here is a snapshot of what Americans had to say in the fall of 2011:

- A shocking 89 percent of Americans said that they distrust government to do the right thing, the highest level of distrust ever recorded.
- Two-thirds of the public said that wealth should be more evenly distributed in the country.
- Forty-three percent of Americans said that they generally agreed with the views of the Occupy Wall Street movement, an amorphous protest movement that began near Wall Street in New York and was spreading to other cities around the country.* A

---

* According to its website, "Occupy Wall Street is a people-powered movement that began on September 17, 2011, in Liberty Square in Manhattan's Financial District, and has spread to over 100 cities in the United States and actions in over 1,500 cities globally. Occupy Wall Street is fighting back against the corrosive power of major

slightly higher percentage, 46 percent, said that the views of the people involved in the Occupy Wall Street movement "generally reflect the views of most Americans."

- Forty-six percent of Americans approved of Barack Obama's handling of his job as president—and an identical 46 percent disapproved of his job performance.
- A mere 9 percent of the public approved of the way Congress was handling its job.
- Even though the presidential primaries would begin in only two months, roughly 80 percent of Republican primary voters said "it was still too early to tell whom they will support."

These are fascinating figures that provided meaningful insight into American opinions one year in advance of a presidential race. Still, one might reasonably ask, How do we know all this? How can we draw such sweeping conclusions about the attitudes of hundreds of millions of adults? And how do we know whether these sweeping conclusions are accurate?

The answer, of course, is that we conduct polls. Or in the example above, the *New York Times* and CBS News can do a poll. (The fact that two competing news organizations would collaborate on a project like this is the first clue that conducting a methodologically sound national poll is not cheap.) I have no doubt that you are familiar with polling results. It may be less obvious that the methodology of polling is just one more form of statistical inference. A poll (or survey) is an inference about the opinions of some population that is based on the views expressed by some sample drawn from that population.

The power of polling stems from the same source as our previous sampling examples: the central limit theorem. If we take a large, represen-

---

banks and multinational corporations over the democratic process, and the role of Wall Street in creating an economic collapse that has caused the greatest recession in generations. The movement is inspired by popular uprisings in Egypt and Tunisia, and aims to expose how the richest 1% of people are writing the rules of an unfair global economy that is foreclosing on our future."

tative sample of American voters (or any other group), we can reasonably assume that our sample will look a lot like the population from which it is drawn. If exactly half of American adults disapprove of gay marriage, then our best guess about the attitudes of a representative sample of 1,000 Americans is that about half of them will disapprove of gay marriage.

Conversely—and more important from the standpoint of polling—if we have a representative sample of 1,000 Americans who feel a certain way, such as the 46 percent who disapprove of President Obama's job performance, then we can infer from that sample that the general population is likely to feel the same way. In fact, we can calculate the probability that our sample results will deviate wildly from the true attitudes of the population. When you read that a poll has a "margin of error" of ± 3 percent, this is really just the same kind of 95 percent confidence interval that we calculated in the last chapter. Our "95 percent confidence" means that if we conducted 100 different polls on samples drawn from the same population, we would expect the answers we get from our sample in 95 of those polls to be within 3 percentage points in one direction or the other of the population's true sentiment. In the context of the job approval question in the *New York Times*/CBS poll, we can be 95 percent confident that the true proportion of all Americans who disapprove of President Obama's job rating lies in the range of 46 percent ± 3 percent, or between 43 percent and 49 percent. If you read the small print on the *New York Times*/CBS poll (as I urge you to do), that's pretty much what it says: "In theory, in 19 cases out of 20, overall results based on such samples will differ by no more than 3 percentage points in either direction from what would have been obtained by seeking to interview all American adults."

One fundamental difference between a poll and other forms of sampling is that the sample statistic we care about will be not a mean (e.g., 187 pounds) but rather a percentage or proportion (e.g., 47 percent of voters, or .47). In other respects, the process is identical. When we have a large, representative sample (the poll), we would expect the proportion of respondents who feel a certain way in the sample (e.g., the 9 percent who think Congress is doing a good job) to be roughly equal to the proportion of all Americans who feel that way. This is no different from assuming that the mean weight for a sample of 1,000 American men should be roughly equal to the mean weight for all American men. Still, we expect some variation in the percentage who approve of Congress from sample to sample, just as we would expect some variation in mean weight as we took different random samples of 1,000 men. If the *New York Times* and CBS had conducted a second poll—asking the same questions to a new sample of 1,000 U.S. adults—it is highly unlikely that the results of the second poll would have been identical to the results of the first. On the other hand, we should not expect the answers from our second sample to diverge widely from the answers given by the first. (To return to a metaphor used earlier, if you taste a spoonful of soup, stir the pot, and then taste again, the two spoonfuls are going to taste similar.) The standard error is what tells us how much dispersion we can expect in our results from sample to sample, which in this case means poll to poll.

The formula for calculating a standard error for a percentage or proportion is slightly different from the formula introduced earlier; the intuition is exactly the same. For any properly drawn random sample, the standard error is equal to $\sqrt{p(1 - p)/n}$, where p is the proportion of respondents expressing a particular view, $(1 - p)$ is the proportion of respondents expressing a different view, and n is the total number of respondents in the sample. You should see that the standard error will fall as the sample size gets larger, since n is in the denominator. The standard error also tends to be smaller when p and $(1 - p)$ are far apart. For example, the standard error will be smaller for a poll in which 95 percent of respondents express a certain view than for a poll in which opinions tend to split 50-50. This is just math, since $(.05)(.95) = .047$, while $(.5)(.5) = .25$; a smaller number in the numerator of the formula leads to a smaller standard error.

As an example, assume that a simple "exit poll" of 500 representative voters on election day finds that 53 percent voted for the Republican candidate; 45 percent of voters voted for the Democrat; and 2 percent supported a third-party candidate. If we use the Republican candidate as our proportion of interest, the standard error for this exit poll would be $\sqrt{(.53)(1 - .53)/500} = \sqrt{(.53)(.47)/500} = \sqrt{.25/500} = \sqrt{.0005} = .02236$.

For simplicity, we'll round the standard error for this exit poll to .02. So far, that's just a number. Let's work through why that number matters. Assume the polls have just closed, and you work for a television network that is keen to declare a winner in the race before the full results are available. You are now the official network data cruncher (having read two-thirds of this book), and your producer wants to know whether it is possible to "call the race" on the basis of this exit poll.

You explain that the answer depends on how confident the network people would like to be in the announcement—or, more specifically, what risk they are willing to take that they will get it wrong. Remember, the standard error gives us a sense of how often we can expect our sample proportion (the exit poll) to lie reasonably close to the true population proportion (the election outcome). We know that roughly 68 percent of the time we can expect the sample proportion—the 53 percent of voters who said they voted for the Republican in this case—to be within one standard error of the true final tally. As a result, you tell your producer "with 68 percent confidence" that your sample, which shows the Republican getting 53 percent of the vote ± 2 percent, or between 51 and 55 percent, has captured the Republican candidate's true tally. Meanwhile, the same exit poll shows that the Democratic candidate has received 45 percent of the vote. If we assume that the vote tally for the Democratic candidate has the same standard error (a simplification that I'll explain in a minute), we can say with 68 percent confidence that the exit poll sample, which shows the Democrat with 45 percent of the vote ± 2 percent, or between 43 and 47 percent, contains the Democrat's true tally. According to this calculation, the Republican is the winner.

The graphics department rushes to do a fancy three-dimensional image that you can flash on the screen for your viewers:

Republican 53%
Democrat 45%
Independent 2%
(Margin of Error 2%)

At first, your producer is impressed and excited, in large part because the above graphic is 3-D, multicolored, and able to spin around on the screen. However, when you explain that roughly 68 times out of 100 your exit poll results will be within one standard error of the true election outcome, your producer, who has twice been sent by the courts to anger management programs, points out the obvious math—32 times out of 100 your exit poll *will not be* within one standard error of the true election outcome. Then what?

You explain that there are two possibilities: (1) the Republican candidate could have received *even more* votes than your poll predicted, in which case you still will have called the election correctly. Or (2) there is a reasonably high probability that the Democratic candidate has received far more votes than your poll has reported, in which case your fancy 3-D, multicolored, spinning graphic will have reported the wrong winner.

Your producer hurls a coffee mug across the room and uses several phrases that violate her probation. She screams, "How can we be [deleted] sure that we have the right [deleted] result?"

Ever the statistics guru, you point out that you cannot be certain of any result until all of the votes are counted. However, you can offer a 95 percent confidence interval instead. In this case, your spinning, 3-D, multicolored graphic will be wrong, on average, only 5 times out of 100.

Your producer lights a cigarette and seems to relax. You decide not to mention the ban on smoking in the workplace, as that turned out disastrously last time. However, you do share some bad news. The only way the station can be more confident of its polling results is by broadening the "margin of error." And when you do that, there is no longer a clear winner in the election. You show your boss the new fancy graphic:

Republican 53%
Democrat 45%
Independent 2%
(Margin of Error 4%)

We know from the central limit theorem that roughly 95 percent of sample proportions will lie within *two standard errors* of the true population proportion (which is 4% in this case). Therefore, if we want to be more confident of our polling results, we have to be less ambitious in what we are predicting. As the above graphic illustrates (without the 3-D and color), at the 95 percent confidence level, the television station can announce that the Republican candidate has earned 53 percent of the vote ± 4 percent, or between 49 and 57 percent of the votes cast. Meanwhile, the Democratic candidate has earned 45 percent ± 4 percent, or between 41 and 49 percent of the votes cast.

And, yes, now you have a new problem. At the 95 percent confidence level, you cannot reject the possibility that the two candidates may be tied with 49 percent of the vote each. This is an inevitable trade-off; the only way to become more certain that your polling results will be consistent with the election outcome *without new data* is to become more timid in your prediction. Think about a nonstatistical context. Suppose you tell a friend that you are "pretty sure" that Thomas Jefferson was the third or fourth president. How can you become more confident of your historical knowledge? By being less specific. You are "absolutely positive" that Thomas Jefferson was one of the first five presidents.

Your producer tells you to order a pizza and prepare to stay at work all night. At that point, statistical good fortune shines upon you. The results of a second exit poll come across your desk with a sample of 2,000 voters. These results show the following: Republican (52 percent); Democrat (45 percent); Independent (3 percent). Your producer is now thoroughly exasperated, since this poll suggests that the gap between the candidates has narrowed, making it even harder for you to call the race in a timely manner. But wait! You point out (heroically) that the sample size (2,000) is four times as large as the sample in the first poll. As a result, the standard error will shrink significantly. The new standard error for the Republican candidate is $\sqrt{.52(.48)/2{,}000}$, which is .01.

If your producer is still comfortable with a 95 percent confidence level, you can declare the Republican candidate the winner. With your new .01 standard error, the 95 percent confidence intervals for the candidates are the following: Republican: 52 ± 2, or between 50 and 54 percent of the votes cast; Democrat: 45 ± 2, or between 43 and 47 percent of the votes cast. There is no longer any overlap between the two confidence intervals. You can predict on air that the Republican candidate is the winner; more than 95 times out of 100 you will be correct.*

But this case is even better than that. The central limit theorem tells us that 99.7 percent of the time a sample proportion will be within three standard errors of the true population proportion. In this election example, our 99.7 percent confidence intervals for the two candidates are the following: Republican, 52 ± 3 percent, or between 49 and 55 percent; Democrat, 45 ± 3 percent, or between 42 and 48 percent. If you report that the Republican candidate has won, there is only a tiny chance that you and your producer will be fired, thanks to your new 2,000-voter sample.

You should see that a bigger sample makes for a shrinking standard error, which is how large national polls can end up with shockingly accurate results. On the other hand, smaller samples obviously make for larger standard errors and therefore a larger confidence interval (or "margin of sampling error," to use the polling lingo). The fine print in the *New York Times*/CBS poll points out that the margin of error for the questions about the Republican primary is 5 percentage points, compared with 3 percentage points for other questions in the poll. Only self-described Republican primary and caucus voters were asked these questions, so the sample size for this subgroup of questions fell to 455 (compared with 1,650 adults for the balance of the poll).

As usual, I've simplified lots of things in this chapter. You might have recognized that in my election example above, the Republican and Democratic candidates should each have their own standard error. Think again about the formula: $SE = \sqrt{p(1-p)/n}$. The size of the sample, n, is

---

* We would expect the Republican candidate's true vote tally to be outside of the confidence interval of the poll roughly 5 percent of the time. In those cases, his true vote tally would be less than 50 percent or greater than 54 percent. However, if he gets more than 54 percent of the vote, your station has not made an error in declaring him the winner. (You've only understated the margin of his victory.) As a result, the probability that your poll will cause you to mistakenly declare the Republican candidate the winner is only 2.5 percent.

the same for both candidates, but p and (1 – p) will be slightly different. In the second exit poll (with the 2,000-voter sample), the standard error for the Republican is $\sqrt{.52(.48)/2{,}000}$ = .01117; for the Democrat, SE = $\sqrt{.45(.55)/2{,}000}$ = .01112. Of course, for all intents and purposes, those two numbers are the same. For that reason, I have adopted a common convention, which is to take the higher standard error of the two and use that for all of the candidates. If anything, this introduces a little extra caution into our confidence intervals.

Many national polls that ask multiple questions will go one step further. In the case of the *New York Times*/CBS poll, the standard error should technically be different for each question, depending on the response. For example, the standard error for the finding that 9 percent of the public approves of the way Congress is handling its job should be lower than the standard error for the question finding that 46 percent of the public approves of the way President Obama has handled his job, since .09 × (.91) is less than .46 × (.54)—.0819 versus .2484. (The intuition behind this formula is explained in a chapter appendix.)

Since it would be both confusing and inconvenient to have a different standard error for each question, polls of this nature will typically assume that the sample proportion for each question is .5 (or 50 percent)—generating the largest possible standard error for any given sample size—and then adopt that standard error to calculate the margin of sampling error for the entire poll.*

When done properly, polls are uncanny instruments. According to Frank Newport, editor in chief of the Gallup Organization, a poll of 1,000 people can offer meaningful and accurate insights into the attitudes of the entire country. Statistically speaking, he's right. But to get those meaningful and accurate results, we have to conduct a proper poll and then interpret the results correctly, both of which are much easier

* The formula for calculating the standard error of a poll that I have introduced here assumes that the poll is conducted on a random sample of the population. Sophisticated polling organizations may deviate from this sampling method, in which case the formula for calculating the standard error will also change slightly. The basic methodology remains the same, however.

said than done. Bad polling results do not typically stem from bad math when calculating the standard errors. Bad polling results typically stem from a biased sample, or bad questions, or both. The mantra "garbage in, garbage out" applies in spades when it comes to sampling public opinion. Below are the key methodological questions one ought to ask when conducting a poll, or when reviewing the work of others.

Is this an accurate sample of the population whose opinions we are trying to measure? Many common data-related challenges were discussed in Chapter 7. Nonetheless, I will point out once again the danger of selection bias, particularly self-selection. Any poll that depends on individuals who select into the sample, such as a radio call-in show or a voluntary Internet survey, will capture only the views of those who make the effort to voice their opinions. These are likely to be the people who feel particularly strongly about an issue, or those who happen to have a lot of free time on their hands. Neither of these groups is likely to be representative of the public at large. I once appeared as a guest on a call-in radio show. One of the callers to the program declared emphatically on air that my views were "so wrong" that he had pulled his car off the highway and found a pay phone in order to call the show and register his dissent. I'd like to think that the listeners who did not pull their cars off the highway to call the show felt differently.

Any method of gathering opinion that systematically excludes some segment of the population is also prone to bias. For example, mobile phones have introduced a host of new methodological complexities. Professional polling organizations go to great lengths to poll a representative sample of the relevant population. The *New York Times*/CBS poll was based on telephone interviews conducted over six days with 1,650 adults, 1,475 of whom said they were registered to vote.

I can only guess at the rest of the methodology, but most professional polls use some variation on the following techniques. To ensure that the adults who pick up the phone are representative of the population, the process starts with probability—a variation on picking marbles out of an urn. A computer randomly selects a set of landline telephone exchanges. (An exchange is an area code plus the first three digits of a phone number.)

By randomly choosing from the 69,000 residential exchanges in the country, each in proportion to its share of all telephone numbers, the survey is likely to get a generally representative geographic distribution of the population. As the small print explains, "The exchanges were chosen so as to ensure that each region of the country was represented in proportion to its share of all telephone numbers." For each exchange selected, the computer added four random digits. As a result, both listed and unlisted numbers will end up on the final list of households to be called. The survey also included a "random dialing of cell phone numbers."

For each number dialed, one adult is designated to be the respondent by a "random procedure," such as asking for the youngest adult who is currently at home. This process has been refined to produce a sample of respondents that resembles the adult population in terms of age and gender. Most important, the interviewer will attempt to make multiple calls at different times of day and evening in order to reach each selected phone number. These repeated attempts—as many as ten or twelve calls to the same number—are an important part of getting an unbiased sample. Obviously it would be cheaper and easier to make random calls to different numbers until a sufficiently large sample of adults have picked up the phone and answered the relevant questions. However, such a sample would be biased toward people who are likely to be home and to answer the phone: the unemployed, the elderly, and so on. That's just fine as long as you're willing to qualify your poll results in the following way: President Obama's approval rating stands at 46 percent among the unemployed, old people, and others who are eager to answer random phone calls.

One indicator of a poll's validity is the response rate: What proportion of respondents who were chosen to be contacted ultimately completed the poll or survey? A low response rate can be a warning sign for potential sampling bias. The more people there are who opt not to answer the poll, or who just can't be reached, the greater the possibility that this large group is different in some material way from those who did answer the questions. Pollsters can test for "nonresponse bias" by analyzing available data on the respondents whom they were not able to contact. Do they live in a particular area? Are they refusing to answer for a particular reason? Are they more likely to be from a particular racial, ethnic, or income group? This kind of analysis can determine whether or not a low response rate will affect the results of the poll.

Have the questions been posed in a way that elicits accurate information on the topic of interest? Soliciting public opinion requires more nuance than measuring test scores or putting respondents on a scale to determine their weight. Survey results can be extremely sensitive to the way a question is asked. Let's take a seemingly simple example: What proportion of Americans support capital punishment? As the chapter title suggests, a solid and consistent majority of Americans approve of the death penalty. According to Gallup, in every year since 2002 over 60 percent of Americans have said they favor the death penalty for a person convicted of murder. The percentage of Americans supporting capital punishment has fluctuated in a relatively narrow range from a high of 70 percent in 2003 to a low of 64 percent at several different points. The polling data are clear: Americans support the death penalty by a wide margin.

Or not. American support for the death penalty plummets *when life imprisonment without parole is offered as an alternative*. A 2006 Gallup poll found that only 47 percent of Americans judged the death penalty as the appropriate penalty for murder, as opposed to 48 percent who preferred life imprisonment.[2] That's not just a statistical factoid to amuse guests at a dinner party; it means that there is no longer majority support for capital punishment when life in prison without parole is a credible alternative. When we solicit public opinion, the phrasing of the question and the choice of language can matter enormously.

Politicians will often exploit this phenomenon by using polls and focus groups to test "words that work." For example, voters are more inclined to support "tax relief" than "tax cuts," even though the two phrases describe the same thing. Similarly, voters are less concerned about "climate change" than they are about "global warming," even though global warming is a form of climate change. Obviously politicians are trying to manipulate voters' responses by choosing nonneutral words. If pollsters are to be considered honest brokers generating legitimate results, they must guard against language that is prone to affect the accuracy of the information collected. Similarly, if answers are to be compared over time—such as how consumers feel about the economy today compared with how they felt a year ago—then the questions eliciting that information over time must be the same, or very similar.

Polling organizations like Gallup will often conduct "split sample testing," in which variations of a question are tested on different samples to gauge how small changes in wording affect respondents' answers. To experts like Gallup's Frank Newport, the answers to every question present meaningful data, even when those answers may appear to be inconsistent.[3] The fact that American attitudes toward capital punishment change dramatically when life without parole is offered as an option tells us something important. The key point, says Newport, is to view any polling result in context. No single question or poll can capture the full depth of public opinion on a complex issue.

Are respondents telling the truth? Polling is like Internet dating: There is a little wiggle room in the veracity of information provided. We know that people shade the truth, particularly when the questions asked are embarrassing or sensitive. Respondents may overstate their income, or inflate the number of times they have sex in a typical month. They may not admit that they do not vote. They may hesitate to express views that are unpopular or socially unacceptable. For all these reasons, even the most carefully designed poll is dependent on the integrity of the respondents' answers.

Election polls depend crucially on sorting those who will vote on Election Day from those who will not. (If we are trying to gauge the likely winner of an election, we do not care about the opinions of anyone who is not going to vote.) Individuals often say they are going to vote because they think that is what pollsters want to hear. Studies that have compared self-reported voting behavior to election records consistently find that one-quarter to one-third of respondents say they voted when in fact they did not.[4] One way to minimize this potential bias is to ask whether a respondent voted in the last election, or in the last several elections. Respondents who have voted consistently in the past are most likely to vote in the future. Similarly, if there are concerns that respondents may be hesitant to express a socially unacceptable answer, such as a negative view of a racial or ethnic group, the question may be phrased in a more subtle way, such as by asking "if people you know" hold such an opinion.

One of the most sensitive surveys of all time was a study conducted by the National Opinion Research Center (NORC) at the University of Chicago called "The Social Organization of Sexuality: Sexual Practices in the United States," which quickly became known as the "Sex Study."[5] The formal description of the study included phrases like "the organization of the behaviors constituting sexual transactions" and "sexual partnering and behavior across the lifecourse." (I'm not even sure what a "lifecourse" is; spell-check says it's not a word.) I'm oversimplifying when I write that the survey sought to document who is doing what to whom—and how often. The purpose of the study, which was published in 1995, was not merely to enlighten us all about the sexual behavior of our neighbors (though that was part of it) but also to gauge how sexual behavior in the United States was likely to affect the spread of HIV/AIDS.

If Americans are hesitant to admit when they don't vote, you can imagine how keen they are to describe their sexual behavior, particularly when it may involve illicit activity, infidelity, or just really weird stuff. The Sex Study methodology was impressive. The research was based on ninety-minute interviews of 3,342 adults chosen to be representative of the U.S. adult population. Nearly 80 percent of the selected respondents completed the survey, leading the authors to conclude that the findings are an accurate reporting of America's sexual behavior (or at least what we were doing in 1995).

Since you've suffered through a chapter on polling methodology, and now nearly an entire book on statistics, you are entitled to a glimpse at what they found (none of which is particularly shocking). As one reviewer noted, "There is much less sexual behavior going on than we might think."[6]

- People generally have sex with others like themselves. Ninety percent of couples shared the same race, religion, social class, and general age group.
- The typical respondent was engaging in sexual activity "a few times a month," though there was wide variation. The number of sexual partners since age eighteen ranged from zero to over 1,000.
- Roughly 5 percent of men and 4 percent of women reported some sexual activity with a same-gender partner.
- Eighty percent of respondents had either one sexual partner in the previous year or none at all.

- Respondents with one sexual partner were happier than those with none or with multiple partners.[7]
- A quarter of the married men and 10 percent of the married women reported having extramarital sexual activity.
- Most people are doing it the old-fashioned way: vaginal intercourse was the most appealing sexual activity for men and women.

One review of the Sex Study made a simple but potent critique: The conclusion that the accuracy of the survey represents the sexual practices of adults in the United States "assumes that respondents to the NORC survey both mirrored the population from which they were drawn and gave accurate answers."[8] That sentence could also be the takeaway for this entire chapter. At first glance, the most suspicious thing about polling is that the opinions of so few can tell us about the opinions of so many. *But that's the easy part.* One of the most basic statistical principles is that a proper sample will look like the population from which it is drawn. The real challenge of polling is twofold: finding and reaching that proper sample; and eliciting information from that representative group in a way that accurately reflects what its members believe.

APPENDIX TO CHAPTER 10

Why is the standard error larger when p (and 1 – p) are close to 50 percent?

Here is the intuition for why the standard error is highest when the proportion answering a particular way (p) is near 50 percent (which, just as a matter of math, means that 1 – p will also be close to 50 percent). Let's imagine that you are conducting two polls in North Dakota. The first poll is designed to measure the mix of Republicans and Democrats in the state. Assume that the true political mix in the North Dakota population is evenly split 50-50 but that your poll finds 60 percent Republicans and 40 percent Democrats. Your results are off by 10 percentage points, which is a large margin. Yet, you have generated this large error without making an unimaginably large data-collecting mistake. You have overcounted the Republicans relative to their true incidence in the population by 20 percent [(60 – 50)/50]. And in so doing, you have also undercounted the Democrats by 20 percent [(40 – 50)/50]. That could happen, even with a decent polling methodology.

Your second poll is designed to measure the fraction of Native Americans in the North Dakota population. Assume that the true proportion of Native Americans in North Dakota is 10 percent while non–Native Americans make up 90 percent of the state population. Now let's discuss how bad your data collecting would have to be in order to produce a poll with a sampling error of 10 percentage points. This could happen in two ways. First, you could find that 0 percent of the population is Native American and 100 percent is non–Native American. Or you could find that 20 percent of the population is Native American and 80 percent is non–Native American. In one case you have missed *all of the Native Americans*; and in the other, you have found *double their true incidence in the population*. These are really bad sampling mistakes. In both cases, your estimate is off by 100 percent: either [(0 – 10)/10] or [(20 – 10)/10]. And if you missed just 20 percent of the Native Americans—the same degree of error that you had in the Republican-Democrat poll—your results would find 8 percent Native Americans and 92 percent non–Native Americans, which is only 2 percentage points from the true split in the population.

When p and 1 – p are close to 50 percent, relatively small sampling errors are magnified into large absolute errors in the outcome of the poll.

When either p or 1 – p is closer to zero, the opposite is true. Even relatively large sampling errors produce small absolute errors in the outcome of the poll.

The same 20 percent sampling error distorted the outcome of the Democrat-Republican poll by 10 percentage points while distorting the Native American poll by only 2 percentage points. Since the standard error in a poll is measured in absolute terms (e.g., ± 5 percent), the formula recognizes that this error is likely to be largest when p and 1 – p are close to 50 percent.

CHAPTER 11

Regression Analysis

The miracle elixir

Can stress on the job kill you? Yes. There is compelling evidence that rigors on the job can lead to premature death, particularly of heart disease. But it's not the kind of stress you are probably imagining. CEOs, who must routinely make massively important decisions that determine the fate of their companies, are at significantly *less* risk than their secretaries, who dutifully answer the phone and perform other tasks as instructed. How can that possibly make sense? It turns out that the most dangerous kind of job stress stems from having "low control" over one's responsibilities. Several studies of thousands of British civil servants (the Whitehall studies) have found that workers who have little control over their jobs—meaning they have minimal say over what tasks are performed or how those tasks are carried out—have a significantly higher mortality rate than other workers in the civil service with more decision-making authority. According to this research, it is not the stress associated with major responsibilities that will kill you; it is the stress associated with being told what to do while having little say in how or when it gets done.

This is not a chapter about job stress, heart disease, or British civil servants. The relevant question regarding the Whitehall studies (and others like them) is how researchers can possibly come to such a conclusion.

Clearly this cannot be a randomized experiment. We cannot arbitrarily assign human beings to different jobs, force them to work in those jobs for many years, and then measure who dies at the highest rate. (Ethical concerns aside, we would presumably wreak havoc on the British civil service by randomly distributing jobs.) Instead, researchers have collected detailed longitudinal data on thousands of individuals in the British civil service; these data can be analyzed to identify meaningful associations, such as the connection between "low control" jobs and coronary heart disease.

A simple association is not enough to conclude that certain kinds of jobs are bad for your health. If we merely observe that low-ranking workers in the British civil service hierarchy have higher rates of heart disease, our results would be confounded by other factors. For example, we would expect low-level workers to have less education than senior officials in the bureaucracy. They may be more likely to smoke (perhaps because of their job frustration). They may have had less healthy childhoods, which diminished their job prospects. Or their lower pay may limit their access to health care. And so on. The point is that any study simply comparing health outcomes across a large group of British workers—or across any other large group—will not really tell us much. Other sources of variation in the data are likely to obscure the relationship that we care about. Is "low job control" really causing heart disease? Or is it some combination of other factors that happen to be shared by people with low job control, in which case we may be completely missing the real public health threat.

Regression analysis is the statistical tool that helps us deal with this challenge. Specifically, regression analysis allows us to quantify the relationship between a particular variable and an outcome that we care about while *controlling for other factors.* In other words, we can isolate the effect of one variable, such as having a certain kind of job, while holding the effects of other variables constant. The Whitehall studies used regression analysis to measure the health impacts of low job control among people who are similar in other ways, such as smoking behavior. (Low-level workers do in fact smoke more than their superiors; this explains a relatively small amount of the variation in heart disease across the Whitehall hierarchy.)

Most of the studies that you read about in the newspaper are based on regression analysis. When researchers conclude that children who spend a lot of time in day care are more prone to behavioral problems in elementary school than children who spend that time at home, the study has not randomly assigned thousands of infants either to day care or to home care with a parent. Nor has the study simply compared the elementary school behavior of children who had different early childhood experiences without recognizing that these populations are likely to be different in other fundamental ways. Different families make different child care decisions *because they are different.* Some households have two parents present; some don't. Some have two parents working; some don't. Some households are wealthier or more educated than others. All of these things affect child care decisions, *and they affect how children in those families will perform in elementary school.* When done properly, regression analysis can help us estimate the effects of day care apart from other things that affect young children: family income, family structure, parental education, and so on.

Now, there are two key phrases in that last sentence. The first is "when done properly." Given adequate data and access to a personal computer, a six-year-old could use a basic statistics program to generate regression results. Personal computing has made the mechanics of regression analysis almost effortless. The problem is that the mechanics of regression analysis are not the hard part; the hard part is determining which variables ought to be considered in the analysis and how that can best be done. Regression analysis is like one of those fancy power tools. It is relatively easy to use, but hard to use well—and potentially dangerous when used improperly.

The second important phrase above is "help us estimate." Our child care study does not give us a "right" answer for the relationship between day care and subsequent school performance. Instead, it quantifies the relationship observed *for a particular group of children over a particular stretch of time.* Can we draw conclusions that might apply to the broader population? Yes, but we will have the same limitations and qualifications as we do with any other kind of inference. First, our sample has to be representative of the population that we care about. A study of 2,000 young children in Sweden will not tell us much about the best policies for early childhood education in rural Mexico. And second, there will be variation from sample to sample. If we do multiple studies of children and child care, each study will produce slightly different findings, even if the methodologies are all sound and similar.

Regression analysis is similar to polling. The good news is that if we have a large representative sample and solid methodology, the relationship we observe for our sample data is not likely to deviate wildly from the true relationship for the whole population. If 10,000 people who exercise three or more times a week have sharply lower rates of cardiovascular disease than 10,000 people who don't exercise (but are similar in all other important respects), then the chances are pretty good that we will see a similar association between exercise and cardiovascular health for the broader population. That's why we do these studies. (The point is not to tell those nonexercisers who are sick at the end of the study that they should have exercised.)

The bad news is that we are not proving definitively that exercise prevents heart disease. We are instead rejecting the null hypothesis that exercise has no association with heart disease, on the basis of some statistical threshold that was chosen before the study was conducted. Specifically, the authors of the study would report that if exercise is unrelated to cardiovascular health, the likelihood of observing such a marked difference in heart disease between the exercisers and nonexercisers in this large sample would be less than 5 in 100, or below some other threshold for statistical significance.

Let's pause for a moment and wave our first giant yellow flag. Suppose that this particular study compared a large group of individuals who play squash regularly with those of an equal-sized group who get no exercise at all. Playing squash does provide a good cardiovascular workout. However, we also know that squash players tend to be affluent enough to belong to clubs with squash courts. Wealthy individuals may have great access to health care, which can also improve cardiovascular health. If our analysis is sloppy, we may attribute health benefits to playing squash when in fact the real benefit comes from being wealthy enough to play squash (in which case playing polo would also be associated with better heart health, even though the horse is doing most of the work).

Or perhaps causality goes the other direction. Could having a healthy heart "cause" exercise? Yes. Individuals who are infirm, particularly those who have some incipient form of heart disease, will find it much harder to exercise. They will certainly be less likely to play squash regularly. Again, if the analysis is sloppy or oversimplified, the claim that exercise is good for your health may simply reflect the fact that people who start out unhealthy find it hard to exercise. In this case, playing squash doesn't make anyone healthier; it merely separates the healthy from the unhealthy.

There are so many potential regression pitfalls that I've devoted the next chapter to the most egregious errors. For now, we'll focus on what can go right. Regression analysis has the amazing capacity to isolate a statistical relationship that we care about, such as that between job control and heart disease, while taking into account other factors that might confuse the relationship.

How exactly does this work? If we know that low-level British civil servants smoke more than their superiors, how can we discern which part of their poor cardiovascular health is due to their low-level jobs, and which part is due to the smoking? These two factors seem inextricably intertwined.

Regression analysis (done properly!) can untangle them. To explain the intuition, I need to begin with the basic idea that underlies all forms of regression analysis—from the simplest statistical relationships to the complex models cobbled together by Nobel Prize winners. At its core, regression analysis seeks to find the "best fit" for a linear relationship between two variables. A simple example is the relationship between height and weight. People who are taller tend to weigh more—though that is obviously not always the case. If we were to plot the heights and weights of a group of graduate students, you might recall what it looked like from Chapter 4:

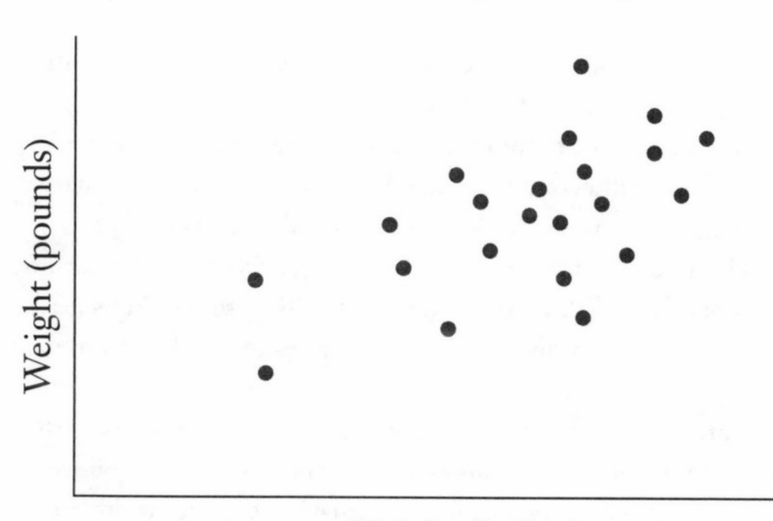

If you were asked to describe the pattern, you might say something along the lines of "Weight seems to increase with height." This is not a terribly insightful or specific statement. Regression analysis enables us to go one step further and "fit a line" that best describes a linear relationship between the two variables.

Many possible lines are broadly consistent with the height and weight data. But how do we know which is the *best* line for these data? In fact, how exactly would we define "best"? Regression analysis typically uses a methodology called ordinary least squares, or OLS. The technical details, including why OLS produces the best fit, will have to be left to a more advanced book. The key point lies in the "least squares" part of the name; OLS fits the line that minimizes the sum of the squared residuals. That's not as awfully complicated as it sounds. Each observation in our height and weight data set has a residual, which is its vertical distance from the regression line, except for those observations that lie directly on the line, for which the residual equals zero. (On the diagram below, the residual is marked for a hypothetical person A.) It should be intuitive that the larger the sum of residuals overall, the worse the fit of the line. The only nonintuitive twist with OLS is that the formula takes the *square* of each residual before adding them all up (which increases the weight given to observations that lie particularly far from the regression line, or the "outliers").

Ordinary least squares "fits" the line that minimizes the sum of the squared residuals, as illustrated below.

Line of Best Fit for Height and Weight

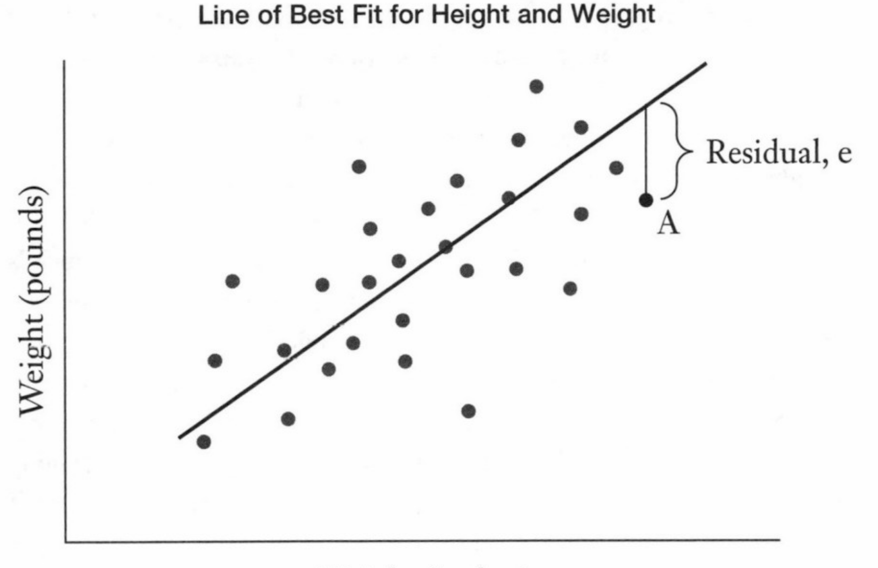

If the technical details have given you a headache, you can be forgiven for just grasping at the bottom line, which is that ordinary least squares gives us the best description of a linear relationship between two variables. The result is not only a line but, as you may recall from high school geometry, an equation describing that line. This is known as the regression equation, and it takes the following form: $y = a + bx$, where y is weight in pounds; a is the y-intercept of the line (the value for y when x = 0); b is the slope of the line; and x is height in inches. The slope of the line we've fitted, b, describes the "best" linear relationship between height and weight for this sample, as defined by ordinary least squares.

The regression line certainly does not describe every observation in the data set perfectly. But it is the best description we can muster for what is clearly a meaningful relationship between height and weight. It also means that every observation can be explained as WEIGHT = a + b(HEIGHT) + e, where e is a "residual" that catches the variation in weight for each individual that is not explained by height. Finally, it means that our best guess for the weight of any person in the data set would be a + b(HEIGHT). Even though most observations do not lie exactly on the regression line, the residual still has an expected value of zero since any person in our sample is just as likely to weigh more than the regression equation predicts as he is to weigh less.

Enough of this theoretical jargon! Let's look at some real height and weight data from the Changing Lives study, though I should first clarify some basic terminology. The variable that is being explained—weight in this case—is known as the dependent variable (because it depends on other factors). The variables that we are using to explain our dependent variable are known as explanatory variables since they explain the outcome that we care about. (Just to make things hard, the explanatory variables are also sometimes called independent variables or control variables.) Let's start by using height to explain weight among the Changing Lives participants; later we'll add other potential explanatory factors.* There are 3,537 adult participants in the Changing Lives study. This is our number of observations, or n. (Sometimes a research paper might note that n = 3,537.) When we run a simple regression on the Changing Lives data with weight as the dependent variable and height as the only explanatory variable, we get the following results:

* You should consider this exercise "fun with data" rather than an authoritative exploration of any of the relationships described in the subsequent regression equations. The purpose here is to provide an intuitive example of how regression analysis works, not to do meaningful research on Americans' weights.

WEIGHT = −135 + (4.5) × HEIGHT IN INCHES

a = −135. This is the y-intercept, which has no particular meaning on its own. (If you interpret it literally, a person who measures zero inches would weigh negative 135 pounds; obviously this is nonsense on several levels.) This figure is also known as the constant, because it is the starting point for calculating the weight of all observations in the study.

b = 4.5. Our estimate for b, 4.5, is known as a regression coefficient, or in statistics jargon, "the coefficient on height," because it gives us the best estimate of the relationship between height and weight among the Changing Lives participants. The regression coefficient has a convenient interpretation: a one-unit increase in the independent variable (height) is associated with an increase of 4.5 units in the dependent variable (weight). For our data sample, this means that a 1-inch increase in height is associated with a 4.5 pound increase in weight. Thus, if we had no other information, our best guess for the weight of a person who is 5 feet 10 inches tall (70 inches) in the Changing Lives study would be −135 + 4.5 (70) = 180 pounds.

This is our payoff, as we have now quantified the best fit for the linear relationship between height and weight for the Changing Lives participants. The same basic tools can be used to explore more complex relationships and more socially significant questions. For any regression coefficient, you will generally be interested in three things: sign, size, and significance.

Sign. The sign (positive or negative) on the coefficient for an independent variable tells us the direction of its association with the dependent variable (the outcome we are trying to explain). In the simple case above, the coefficient on height is positive. Taller people tend to weigh more. Some relationships will work in the other direction. I would expect the association between exercise and weight to be negative. If the Changing Lives study included data on something like "miles run per month," I am fairly certain that the coefficient on "miles run" would be negative. Running more is associated with weighing less.

Size. How big is the observed effect between the independent variable and the dependent variable? Is it of a magnitude that matters? In this case, every one inch in height is associated with 4.5 pounds, which is a sizable percentage of a typical person's body weight. In an explanation of why some people weigh more than others, height is clearly an important factor. In other studies, we may find an explanatory variable that has a statistically significant impact on our outcome of interest—meaning that the observed effect is not likely to be a product of chance—but that effect may be so small as to be trivial or *socially insignificant.* For example, suppose that we are examining determinants of income. Why do some people make more money than others? The explanatory variables are likely to be things like education, years of work experience, and so on. In a large data set, researchers might also find that people with whiter teeth earn $86 more per year than other workers, ceteris paribus. ("Ceteris paribus" comes from the Latin meaning "other things being equal.") The positive and statistically significant coefficient on the "white teeth" variable assumes that the individuals being compared are similar in other respects: same education, same work experience, and so on. (I will explain in a moment how we pull off this tantalizing feat.) Our statistical analysis has demonstrated that whiter teeth are associated with $86 in additional annual income per year and that this finding is not likely to be a mere coincidence. This means (1) we've rejected the null hypothesis that really white teeth have no association with income with a high degree of confidence; and (2) if we analyze other data samples, we are likely to find a similar relationship between good-looking teeth and higher income.

So what? We've found a statistically significant result, but not one that is particularly meaningful. To begin with, $86 per year is not a life-changing sum of money. From a public policy standpoint, $86 is also probably less than it would cost to whiten an individual's teeth every year, so we can't even recommend that young workers make such an investment. And, although I'm getting a chapter ahead of myself, I'd also be worried about some serious methodological problems. For example, having perfect teeth may be associated with other personality traits that explain the earnings advantage; the earnings effect may be caused by the kind of people who care about their teeth, not the teeth themselves. For now, the point is that we should take note of the size of the association that we observe between the explanatory variable and our outcome of interest.

Significance. Is the observed result an aberration based on a quirky sample of data, or does it reflect a meaningful association that is likely to be observed for the population as a whole? This is the same basic question that we have been asking for the last several chapters. In the context of height and weight, do we think that we would observe a similar positive association in other samples that are representative of the population? To answer this question, we use the basic tools of inference that have already been introduced. Our regression coefficient is based on an observed relationship between height and weight for a particular sample of data. If we were to test another large sample of data, we would almost certainly get a slightly different association between height and weight and therefore a different coefficient. The relationship between height and weight observed in the Whitehall data (the British civil servants) is likely to be different from the relationship observed between height and weight for the participants in the Changing Lives study. However, we know from the central limit theorem that the mean for a large, properly drawn sample will not typically deviate wildly from the mean for the population as a whole. Similarly, we can assume that the observed relationship between variables like height and weight will not typically bounce around wildly from sample to sample, assuming that these samples are large and properly drawn from the same population.

Think about the intuition: It's highly unlikely (though still possible) that we would find that every inch of height is associated with 4.5 additional pounds among the Changing Lives participants but that there is no association between height and weight in a different representative sample of 3,000 adult Americans.

This should give you the first inkling of how we'll test whether our regression results are statistically significant or not. As with polling and other forms of inference, we can calculate a standard error for the regression coefficient. The standard error is a measure of the likely dispersion we would observe in the coefficient if we were to conduct the regression analysis on repeated samples drawn from the same population. If we were to measure and weigh a different sample of 3,000 Americans, we might find in the subsequent analysis that each inch of height is associated with 4.3 pounds. If we did it again for another sample of 3,000 Americans, we might find that each inch is associated with 5.2 pounds. Once again, the normal distribution is our friend. For large samples of data, such as our Changing Lives data set, we can assume that our various coefficients will be distributed normally around the "true" association between height and weight in the American adult population. On that assumption, we can calculate a standard error for the regression coefficient that gives us a sense of how much dispersion we should expect in the coefficients from sample to sample. I will not delve into the formula for calculating the standard error here, both because it will take us off in a direction that involves a lot of math and because all basic statistical packages will calculate it for you.

However, I must warn that when we are working with a small sample of data—such as a group of 20 adults rather than the 3,000+ persons in the Changing Lives study—the normal distribution is no longer willing to be our friend. Specifically, if we repeatedly conduct regression analysis on different small samples, we can no longer assume that our various coefficients will be distributed normally around the "true" association between height and weight in the American adult population. Instead, our coefficients will still be distributed around the "true" association between height and weight for the American adult population in what is known as a t-distribution. (Basically the t-distribution is more dispersed than the normal distribution and therefore has "fatter tails.") Nothing else changes; any basic statistical software package will easily manage the additional complexity associated with using the t-distributions. For this reason, the t-distribution will be explained in greater detail in the chapter appendix.

Sticking with large samples for now (and the normal distribution), the most important thing to understand is why the standard error matters. As with polling and other forms of inference, we expect that more than half of our observed regression coefficients will lie within one standard error of the true population parameter.* Roughly 95 percent will lie within two standard errors. And so on. With that, we're just about

* "Parameter" is a fancy term for any statistic that describes a characteristic of some population; the mean weight for all adult men is a parameter of that population. So is the standard deviation. In the example here, the true association between height and weight for the population is a parameter of that population.

home, because now we can do a little hypothesis testing. (Seriously, did you think you were already done with hypothesis testing?) Once we have a coefficient and a standard error, we can test the null hypothesis that there is in fact no relationship between the explanatory variable and the dependent variable (meaning that the true association between the two variables in the population is zero).

In our simple height and weight example, we can test how likely it is that we would find in our Changing Lives sample that every inch of height is associated with 4.5 pounds if there is really no association between height and weight in the general population. I've run the regression by using a basic statistics program; the standard error on the height coefficient is .13. This means that if we were to do this analysis repeatedly—say with 100 different samples—then we would expect our observed regression coefficient to be within two standard errors of the true population parameter roughly 95 times out of 100.

We can therefore express our results in two different but related ways. First, we can build a 95 percent confidence interval. We can say that 95 times out of 100, we expect our confidence interval, which is 4.5 ± .26, to contain the true population parameter. This is the range between 4.24 and 4.76. A basic statistics package will calculate this interval as well. Second, we can see that our 95 percent confidence interval for the true association between height and weight does not include zero. Thus, we can reject the null hypothesis that there is *no association* between height and weight for the general population at the 95 percent confidence level. This result can also be expressed as being statistically significant at the .05 level; there is only a 5 percent chance that we are wrongly rejecting the null hypothesis.

In fact, our results are even more extreme than that. The standard error (.13) is extremely low relative to the size of the coefficient (4.5). One rough rule of thumb is that the coefficient is likely to be statistically significant when the coefficient is at least twice the size of the standard error.* A statistics package also calculates a p-value, which is .000 in this

---

* When the null hypothesis is that a regression coefficient is zero (as is most often the case), the ratio of the observed regression coefficient to the standard error is known as the *t-statistic*. This will also be explained in the chapter appendix.

case, meaning that there is essentially zero chance of getting an outcome as extreme as what we've observed (or more so) if there is no true association between height and weight in the general population. Remember, we have not *proved* that taller people weigh more in the general population; we have merely shown that our results for the Changing Lives sample would be highly anomalous if that were not the case.

Our basic regression analysis produces one other statistic of note: the R^2, which is a measure of the total amount of variation explained by the regression equation. We know that we have a broad variation in weight in our Changing Lives sample. Many of the persons in the sample weigh more than the mean for the group overall; many weigh less. The R^2 tells us how much of that variation around the mean is associated with differences in height alone. The answer in our case is .25, or 25 percent. The more significant point may be that 75 percent of the variation in weight for our sample remains unexplained. There are clearly factors other than height that might help us understand the weights of the Changing Lives participants. This is where things get more interesting.

I'll admit that I began this chapter by selling regression analysis as the miracle elixir of social science research. So far all I've done is use a statistics package and an impressive data set to demonstrate that tall people tend to weigh more than short people. A short trip to a shopping mall would probably have convinced you of the same thing. Now that you understand the basics, we can unleash the real power of regression analysis. It's time to take off the training wheels!

As I've promised, regression analysis allows us to unravel complex relationships in which multiple factors affect some outcome that we care about, such as income, or test scores, or heart disease. When we include multiple variables in the regression equation, the analysis gives us an estimate of the linear association between *each* explanatory variable and the dependent variable while holding other dependent variables constant, or "controlling for" these other factors. Let's stick with weight for a while. We've found an association between height and weight; we know there are other factors that can help to explain weight (age, sex, diet, exercise, and so on). Regression analysis (often called multiple regression analysis when more than one explanatory variable is involved, or multivariate regression analysis) will give us a coefficient for each explanatory variable included in the regression equation. In other words, *among people who are the same sex and height*, what is the relationship between age and weight? Once we have more than one explanatory variable, we can no longer plot these data in two dimensions. (Try to imagine a graph that represents the weight, sex, height, and age of each participant in the Changing Lives study.) Yet the basic methodology is the same as in our simple height and weight example. As we add explanatory variables, a statistical package will calculate the regression coefficients that minimize the total sum of the squared residuals for the regression equation.

Let's work with the Changing Lives data for now; then I'll go back and give an intuitive explanation for how this statistical parting of the Red Sea could possibly work. We can start by adding one more variable to the equation that explains the weights of the Changing Lives participants: age. When we run the regression including both height and age as explanatory variables for weight, here is what we get.

WEIGHT = –145 + 4.6 × (HEIGHT IN INCHES)
+ .1 × (AGE IN YEARS)

The coefficient on age is .1. That can be interpreted to mean that every additional year in age is associated with .1 additional pounds in weight, *holding height constant*. For any group of people who are the same height, on average those who are ten years older will weigh one pound more. This is not a huge effect, but it's consistent with what we tend to see in life. The coefficient is significant at the .05 level.

You may have noticed that the coefficient on height has increased slightly. Once age is in our regression, we have a more refined understanding of the relationship between height and weight. Among people who are the same age in our sample, or "holding age constant," every additional inch in height is associated with 4.6 pounds in weight.

Let's add one more variable: sex. This will be slightly different because sex can only take on two possibilities, male or female. How does one put M or F into a regression? The answer is that we use what is called a binary variable, or dummy variable. In our data set, we enter a 1 for those participants who are female and a 0 for those who are male. (This is not meant to be a value judgment.) The sex coefficient can then be interpreted as the effect on weight of being female, ceteris paribus. The coefficient is –4.8, which is not surprising. We can interpret that to mean that for individuals who are the same height and age, women typically weigh 4.8 pounds less than men. Now we can begin to see some of the power of multiple regression analysis. We know that women tend to be shorter than men, but our coefficient takes this into account since we have already controlled for height. What we have isolated here is the effect of being female. The new regression becomes:

WEIGHT = –118 + 4.3 × (HEIGHT IN INCHES)
+ .12 (AGE IN YEARS) – 4.8 (IF SEX IS FEMALE)

Our best estimate of the weight of a fifty-three-year-old woman who is 5 feet 5 inches is: –118 + 4.3 (65) + .12 (53) – 4.8 = 163 pounds.

And our best guess for a thirty-five-year-old male who is 6 feet 3 inches is –118 + 4.3 (75) + .12 (35) = 209 pounds. We skip the last term in our regression result (–4.8) since this person is not female.

Now we can start to test things that are more interesting and less predictable. What about education? How might that affect weight? I would hypothesize that better-educated individuals are more health conscious and therefore will weigh less, ceteris paribus. We also haven't tested any measure of exercise; I assume that, holding other factors constant, the people in the sample who get more exercise will weigh less.

What about poverty? Does being low-income in America have effects that manifest themselves in weight? The Changing Lives study asks whether the participants are receiving food stamps, which is a good measure of poverty in America. Finally, I'm interested in race. We know that people of color have different life experiences in the United States *because of their race*. There are cultural and residential factors associated with race in America that have implications for weight. Many cities are still characterized by a high degree of racial segregation; African Americans might be more likely than other residents to live in "food deserts," which are areas with limited access to grocery stores that carry fruits, vegetables, and other fresh produce.

We can use regression analysis to separate out the independent effect of each of the potential explanatory factors described above. For example, we can isolate the association between race and weight, holding constant other socioeconomic factors like educational background and poverty. *Among people who are high school graduates and eligible for food stamps, what is the statistical association between weight and being black?*

At this point, our regression equation is so long that it would be cumbersome to print the results in their entirety here. Academic papers typically insert large tables that summarize the results of various regression equations. I have included a table with the complete results of this regression equation in the appendix to this chapter. In the meantime, here are the highlights of what happens when we add education, exercise, poverty (as measured by receiving food stamps), and race to our equation.

All of our original variables (height, age, and sex) are still significant. The coefficients change little as we add explanatory variables. All of our new variables are statistically significant at the .05 level. The R^2 on the regression has climbed from .25 to .29. (Remember, an R^2 of zero means that our regression equation does no better than the mean at predicting the weight of any individual in the sample; an R^2 of 1 means that the regression equation perfectly predicts the weight of every person in the sample.) A lot of the variation in weight across individuals remains unexplained.

Education turns out to be negatively associated with weight, as I had hypothesized. Among participants in the Changing Lives study, each year of education is associated with –1.3 pounds.

Not surprisingly, exercise is also negatively associated with weight. The Changing Lives study includes an index that evaluates each participant in the study on his or her level of physical activity. Those individuals who are in the bottom quintile of physical activity weigh, on average, 4.5 pounds more than other adults in the sample, ceteris paribus. Those in the bottom quintile for physical activity weigh, on average, nearly 9 pounds more than adults in the top quintile for physical activity.

Individuals receiving food stamps (the proxy for poverty in this regression) are heavier than other adults. Food stamp recipients weigh an average of 5.6 pounds more than other Changing Lives participants, ceteris paribus.

The race variable turns out to be particularly interesting. Even after one controls for all the other variables described up to this point, race still matters a lot when it comes to explaining weight. The non-Hispanic black adults in the Changing Lives sample weigh, on average, roughly 10 pounds more than the other adults in the sample. Ten pounds is a lot of weight, both in absolute terms and compared with the effects of the other explanatory variables in the regression equation. This is not a quirk of the data. The p-value on the dummy variable for non-Hispanic blacks is .000 and the 95 percent confidence interval stretches from 7.7 pounds to 16.1 pounds.

What is going on? The honest answer is that I have no idea. Let me reiterate a point that was buried earlier in a footnote: I'm just playing around with data here to illustrate how regression analysis works. The analytics presented here are to true academic research what street hockey is to the NHL. If this were a real research project, there would be weeks or months of follow-on analysis to probe this finding. What I can say is that I have demonstrated why multiple regression analysis is the best tool we have for finding meaningful patterns in large, complex data sets. We started with a ridiculously banal exercise: quantifying the relationship between height and weight. Before long, we were knee-deep in issues with real social significance.

In that vein, I can offer you a real study that used regression analysis to probe a socially significant issue: gender discrimination in the workplace. The curious thing about discrimination is that it's hard to observe directly. No employer ever states explicitly that someone is being paid less because of his or her race or gender or that someone has not been hired for discriminatory reasons (which would presumably leave the person in a different job with a lower salary). Instead, what we observe are gaps in pay by race and gender that may be the result of discrimination: whites earn more than blacks; men earn more than women; and so on. The methodological challenge is that these observed gaps may also be the result of underlying differences in workers that have nothing to do with workplace discrimination, such as the fact that women tend to choose more part-time work. How much of the wage gap is due to factors associated with productivity on the job, and how much of the gap, if any, is due to labor force discrimination? No one can claim that this is a trivial question.

Regression analysis can help us answer it. However, our methodology will be slightly more roundabout than it was with our analysis explaining weight. Since we cannot measure discrimination directly, we will examine other factors that traditionally explain wages, such as education, experience, occupational field, and so on. The case for discrimination is circumstantial: If a significant wage gap remains after controlling for other factors that typically explain wages, then discrimination is a likely culprit. The larger the unexplained portion of any wage gap, the more suspicious we should be. As an example, let's look at a paper by three economists examining the wage trajectories of a sample of roughly 2,500 men and women who graduated with MBAs from the Booth School of Business at the University of Chicago.[1] Upon graduation, male and female graduates have very similar average starting salaries: $130,000 for men and $115,000 for women. After ten years in the workforce, however, a huge gap has opened up; women on average are earning a striking 45 percent less than their male classmates: $243,000 versus $442,000. In a broader sample of more than 18,000 MBA graduates who entered the workforce between 1990 and 2006, being female is associated with 29 percent lower earnings. What is happening to women once they enter the labor force?

According to the authors of the study (Marianne Bertrand of the Booth School of Business and Claudia Goldin and Lawrence Katz of Harvard), discrimination is *not* a likely explanation for most of the gap. The gender wage gap fades away as the authors add more explanatory variables to the analysis. For example, men take more finance classes in the MBA program and graduate with higher grade point averages. When these data are included as control variables in the regression equation, the unexplained portion of the gap in male-female earnings drops to 19 percent. When variables are added to the equation to account for post-MBA work experience, particularly spells out of the labor force, the unexplained portion of the male-female wage gap drops to 9 percent. And when explanatory variables are added for other work characteristics, such as employer type and hours worked, the unexplained portion of the gender wage gap falls to under 4 percent.

For workers who have been in the labor force more than ten years, the authors can ultimately explain all but 1 percent of the gender wage gap with factors unrelated to discrimination on the job.* They conclude, "We identify three proximate reasons for the large and rising gender gap in earnings: differences in training prior to MBA graduation; differences in career interruptions; and differences in weekly hours. These three determinants can explain the bulk of gender differences across the years following MBA completion."

I hope that I've convinced you of the value of multiple regression analysis, particularly the research insights that stem from being able to isolate the effect of one explanatory variable while controlling for other confounding factors. I have not yet provided an intuitive explanation for how this statistical "miracle elixir" works. When we use regression analysis to evaluate the relationship between education and weight, ceteris paribus, how does a statistical package control for factors like height, sex, age, and income when we know that our Changing Lives participants are *not* identical in these other respects?

To get your mind around how we can isolate the effect on weight of a single variable, say, education, imagine the following situation. Assume that all of the Changing Lives participants are convened in one place—say, Framingham, Massachusetts. Now assume that the men and women are separated. And then assume that both the men and the women are further divided by height. There will be a room of six-foot tall men. Next door, there will be a room of 6-foot 1-inch men, and so on for both genders. If we have enough participants in our study, we can further subdivide each of those rooms by income. Eventually we will have lots of rooms, each of which contains individuals who are identical in all respects *except for education and weight, which are the two variables we care about*. There would be a room of forty-five-year-old 5-foot 5-inch men

* Broader discriminatory forces in society may affect the careers that women choose or the fact that they are more likely than men to interrupt their careers to take care of children. However, these important issues are distinct from the narrower question of whether women are being paid less than men to do the same jobs.

who earn $30,000 to $40,000 a year. Next door would be all the forty-five-year-old 5-foot 5-inch women who earn $30,000 to $40,000 a year. And so on (and on and on).

There will still be some variation in weight in each room; people who are the same sex and height and have the same income will still weigh different amounts—though presumably there will be much less variation in weight in each room than there is for the overall sample. Our goal now is to see how much of the remaining variation in weight in each room can be explained by education. In other words, what is the best linear relationship between education and weight in each room?

The final challenge, however, is that we do not want different coefficients in each "room." The whole point of this exercise is to calculate a single coefficient that best expresses the relationship between education and weight for the entire sample, while holding other factors constant. What we would like to calculate is the single coefficient for education that we can use *in every room* to minimize the sum of the squared residuals for all of the rooms combined. What coefficient for education minimizes the square of the unexplained weight for every individual across all the rooms? That becomes our regression coefficient because it is the best explanation of the linear relationship between education and weight for this sample when we hold sex, height, and income constant.

As an aside, you can see why large data sets are so useful. They allow us to control for many factors while still having many observations in each "room." Obviously a computer can do all of this in a split second without herding thousands of people into different rooms.

Let's finish the chapter where we started, with the connection between stress on the job and coronary heart disease. The Whitehall studies of British civil servants sought to measure the association between grade of employment and death from coronary heart disease over the ensuing years. One of the early studies followed 17,530 civil servants for seven and a half years.[2] The authors concluded, "Men in the lower employment grades were shorter, heavier for their height, had higher blood pressure, higher plasma glucose, smoked more, and reported less leisure-time physical activity than men in the higher grades. Yet when allowance was made for the influence on mortality of all of these factors plus plasma cholesterol, the inverse association between grade of employment and [coronary heart disease] mortality was still strong." The "allowance" they refer to for these other known risk factors is done by means of regression analysis.* The study demonstrates that holding other health factors constant (including height, which is a decent proxy for early childhood health and nutrition), working in a "low grade" job can literally kill you.

Skepticism is always a good first response. I wrote at the outset of the chapter that "low-control" jobs are bad for your health. That may or may not be synonymous with being low on the administrative totem pole. A follow-up study using a second sample of 10,308 British civil servants sought to drill down on this distinction.[3] The workers were once again divided into administrative grades—high, intermediate, and low—only this time the participants were also given a fifteen-item questionnaire that evaluated their level of "decision latitude or control." These included questions such as "Do you have a choice in deciding how you do your job?" and categorical responses (ranging from "never" to "often") to statements such as "I can decide when to take a break." The researchers found that the "low-control" workers were at significantly higher risk of developing coronary heart disease over the course of the study than "high-control" workers. Yet researchers also found that workers with

---

* These studies differ slightly from the regression equations introduced earlier in the chapter. The outcome of interest, or dependent variable, is binary in these studies. A participant either has some kind of heart-related health problem during the period of study or he does not. As a result, the researchers use a tool called multivariate logistic regression. The basic idea is the same as the ordinary least squares models described in this chapter. Each coefficient expresses the effect of a particular explanatory variable on the dependent variable while holding the effects of other variables in the model constant. The key difference is that the variables in the equation all affect the *likelihood* that some event happens, such as having a heart attack during the period of study. In this study, for example, workers in the low control group are 1.99 times as likely to have "any coronary event" over the period of study as workers in the high control group after controlling for other coronary risk factors.

rigorous job demands were at no greater risk of developing heart disease, nor were workers who reported low levels of social support on the job. Lack of control seems to be the killer, literally.

The Whitehall studies have two characteristics typically associated with strong research. First, the results have been replicated elsewhere. In the public health literature, the "low-control" idea evolved into a term known as "job strain," which characterizes jobs with "high psychological workload demands" and "low decision latitude." Between 1981 and 1993, thirty-six studies were published on the subject; most found a significant positive association between job strain and heart disease.[4]

Second, researchers sought and found corroborating biological evidence to explain the mechanism by which this particular kind of stress on the job causes poor health. Work conditions that involve rigorous demands but low control can cause physiological responses (such as the release of stress-related hormones) that increase the risk of heart disease over the long run. Even animal research plays a role; low-status monkeys and baboons (who bear some resemblance to civil servants at the bottom of the authority chain) have physiological differences from their high-status peers that put them at greater cardiovascular risk.[5]

All else equal, it's better not to be a low-status baboon, which is a point I try to make to my children as often as possible, particularly my son. The larger message here is that regression analysis is arguably the most important tool that researchers have for finding meaningful patterns in large data sets. We typically cannot do controlled experiments to learn about job discrimination or factors that cause heart disease. Our insights into these socially significant issues and many others come from the statistical tools covered in this chapter. In fact, it would not be an exaggeration to say that a high proportion of all important research done in the social sciences over the past half century (particularly since the advent of cheap computing power) draws on regression analysis.

Regression analysis supersizes the scientific method; we are healthier, safer, and better informed as a result.

So what could possibly go wrong with this powerful and impressive tool? Read on.

APPENDIX TO CHAPTER 11

The t-distribution

Life gets a little trickier when we are doing our regression analysis (or other forms of statistical inference) with a small sample of data. Suppose we were analyzing the relationship between weight and height on the basis of a sample of only 25 adults, rather than using a huge data set like the Changing Lives study. Logic suggests that we should be less confident about generalizing our results to the entire adult population from a sample of 25 than from a sample of 3,000. One of the themes throughout the book has been that smaller samples tend to generate more dispersion in outcomes. Our sample of 25 will still give us meaningful information, as would a sample of 5 or 10—but how meaningful?

The t-distribution answers that question. If we analyze the association between height and weight for repeated samples of 25 adults, we can no longer assume that the various coefficients we get for height will be distributed normally around the "true" coefficient for height in the adult population. They will still be distributed around the true coefficient for the whole population, but *the shape of that distribution will not be our familiar bell-shaped normal curve*. Instead, we have to assume that repeated samples of just 25 will produce more dispersion around the true population coefficient—and therefore a distribution with "fatter tails." And repeated samples of 10 will produce even more dispersion than that—and therefore even fatter tails. The t-distribution is actually a series, or "family," of probability density functions that vary according to the size of our sample. Specifically, the more data we have in our sample, the more "degrees of freedom" we have when determining the appropriate distribution against which to evaluate our results. In a more advanced class, you will learn exactly how to calculate degrees of freedom; for our purposes, they are roughly equal to the number of observations in our sample. For instance, a basic regression analysis with a sample of 10 and a single explanatory variable has 9 degrees of freedom. The more degrees of freedom we have, the more confident we can be that our sample represents the true population, and the "tighter" our distribution will be, as the following diagram illustrates.

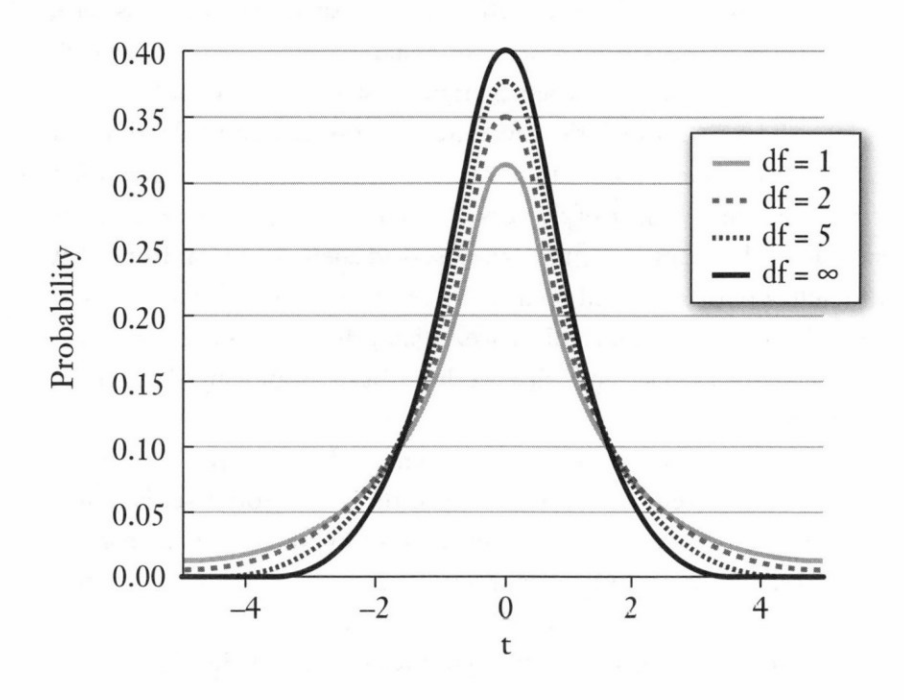

When the number of degrees of freedom gets large, the t-distribution converges to the normal distribution. That's why when we are working with large data sets, we can use the normal distribution for our assorted calculations.

The t-distribution merely adds nuance to the same process of statistical inference that we have been using throughout the book. We are still formulating a null hypothesis and then testing it against some observed data. If the data we observe would be highly unlikely if the null hypothesis were true, then we reject the null hypothesis. The only thing that changes with the t-distribution is the underlying probabilities for evaluating the observed outcomes. The "fatter" the tail in a particular probability distribution (e.g., the t-distribution for eight degrees of freedom), the more dispersion we would expect in our observed data just as a matter of chance, and therefore the less confident we can be in rejecting our null hypothesis.

For example, suppose we are running a regression equation, and the null hypothesis is that the coefficient on a particular variable is zero. Once we get the regression results, we would calculate a t-statistic, which is the ratio of the observed coefficient to the standard error for that coefficient.* This t-statistic is then evaluated against whatever t-distribution is appropriate for the size of the data sample (since this is largely what determines the number of degrees of freedom). When the t-statistic is sufficiently large, meaning that our observed coefficient is far from what the null hypothesis would predict, we can reject the null hypothesis at some level of statistical significance. Again, this is the same basic process of statistical inference that we have been employing throughout the book.

The fewer the degrees of freedom (and therefore the "fatter" the tails of the relevant t-distribution), the higher the t-statistic will have to be in order for us to reject the null hypothesis at some given level of significance. In the hypothetical regression example described above, if we had four degrees of freedom, we would need a t-statistic of at least 2.13 to reject the null hypothesis at the .05 level (in a one-tailed test).

However, if we have 20,000 degrees of freedom (which essentially allows us to use the normal distribution), we would need only a t-statistic of 1.65 to reject the null hypothesis at the .05 level in the same one-tailed test.

---

* The more general formula for calculating a t-statistic is the following:

$$t_b = \frac{b - b_o}{SE_b}$$

where b is the observed coefficient, b_o is the null hypothesis for that coefficient, and SE_b is the standard error for the observed coefficient b.

Regression Equation for Weight

Variable	Coefficient	Standard Error	t-statistic	p-value (two-tailed test)	95% Confidence Interval
Height	4.4	.2	21.4	.000	4.0 to 4.8
Age	.08	.03	2.2	.026	.01 to .2
Sex	–5.7	1.7	–3.4	.001	–9.0 to –2.4
Years of Educational Attainment	–.7	.2	–3.5	.000	–1.1 to –.3
Bottom Quintile of Physical Activity	3.7	1.4	2.6	.009	.9 to 6.5
Dummy for Receiving Food Stamps	5.6	2.1	2.7	.007	1.5 to 9.7
Non-Hispanic Black	9.7	1.3	7.2	.000	7.0 to 12.3
Intercept	–117				

CHAPTER 12

Common Regression Mistakes

The mandatory warning label

Here is one of the most important things to remember when doing research that involves regression analysis: Try not to kill anyone. You can even put a little Post-it note on your computer monitor: "Do not kill people with your research." Because some very smart people have inadvertently violated that rule.

Beginning in the 1990s, the medical establishment coalesced around the idea that older women should take estrogen supplements to protect against heart disease, osteoporosis, and other conditions associated with menopause.[1] By 2001, some 15 million women were being prescribed estrogen in the belief that it would make them healthier. Why? Because research at the time—using the basic methodology laid out in the last chapter—suggested this was a sensible medical strategy. In particular, a longitudinal study of 122,000 women (the Nurses' Health Study) found a negative association between estrogen supplements and heart attacks. Women taking estrogen had one-third as many heart attacks as women who were not taking estrogen. This was not a couple of teenagers using dad's computer to check out pornography and run regression equations. The Nurses' Health Study is run by the Harvard Medical School and the Harvard School of Public Health.

Meanwhile, scientists and physicians offered a medical theory for why hormone supplements might be beneficial for female health. A woman's ovaries produce less estrogen as she ages; if estrogen is important to the body, then making up for this deficit in old age could be protective of a woman's long-term health. Hence the name of the treatment: hormone replacement therapy. Some researchers even began to suggest that older men should receive an estrogen boost.[2]

And then, while millions of women were being prescribed hormone replacement therapy, estrogen was subjected to the most rigorous form of scientific scrutiny: clinical trials. Rather than searching a large data set like the Nurses' Health Study for statistical associations that may or may not be causal, a clinical trial consists of a controlled experiment. One sample is given a treatment, such as hormone replacement; another sample is given a placebo. Clinical trials showed that women taking estrogen had a higher incidence of heart disease, stroke, blood clots, breast cancer, and other adverse health outcomes. Estrogen supplements did have some benefits, but those benefits were far outweighed by other risks. Beginning in 2002, doctors were advised not to prescribe estrogen for their aging female patients. The *New York Times Magazine* asked a delicate but socially significant question: How many women died prematurely or suffered strokes or breast cancer because they were taking a pill that their doctors had prescribed to keep them healthy?

The answer: "A reasonable estimate would be tens of thousands."[3]

Regression analysis is the hydrogen bomb of the statistics arsenal. Every person with a personal computer and a large data set can be a researcher in his or her own home or cubicle. What could possibly go wrong? All kinds of things. Regression analysis provides precise answers to complicated questions. These answers may or may not be accurate. In the wrong hands, regression analysis will yield results that are misleading or just plain wrong. And, as the estrogen example illustrates, *even in the right hands* this powerful statistical tool can send us speeding dangerously in the wrong direction. The balance of this chapter will explain the most common regression "mistakes." I put "mistakes" in quotation marks, because, as with all other kinds of statistical analysis, clever people can knowingly exploit these methodological points to nefarious ends.

Here is a "Top Seven" list of the most common abuses of an otherwise extraordinary tool.

Using regression to analyze a nonlinear relationship.* Have you ever read the warning label on a hair dryer—the part that cautions, Do Not Use in the Bath Tub? And you think to yourself, "What kind of moron uses a hair dryer in the bath tub?" *It's an electrical appliance; you don't use electrical appliances around water.* They're not designed for that. If regression analysis had a similar warning label, it would say, Do Not Use When There Is Not a Linear Association between the Variables That You Are Analyzing. Remember, a regression coefficient describes the slope of the "line of best fit" for the data; a line that is not straight will have a different slope in different places. As an example, consider the following hypothetical relationship between the number of golf lessons that I take during a month (an explanatory variable) and my average score for an eighteen-hole round during that month (the dependent variable). As you can see from the scatter plot, there is no consistent *linear* relationship.

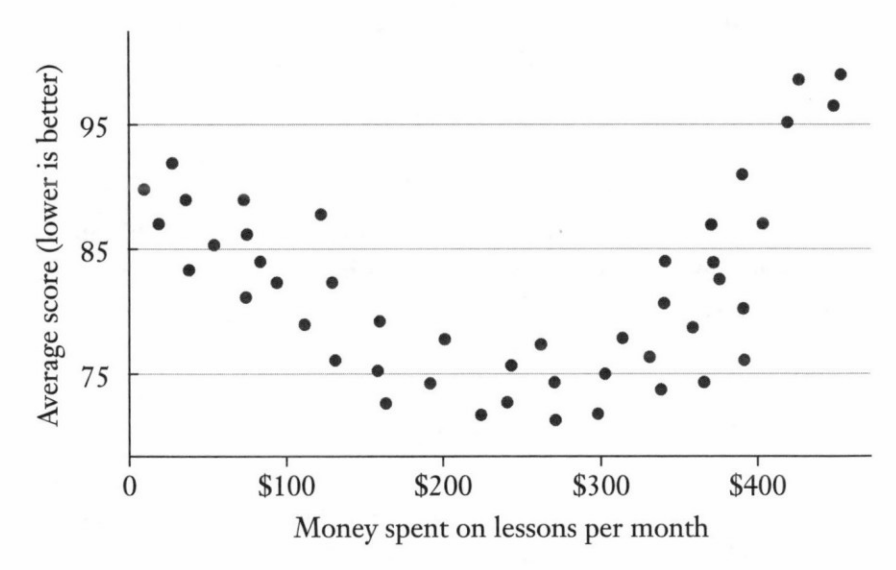

* There are more sophisticated methods that can be used to adapt regression analysis for use with nonlinear data. Before using those tools, however, you need to appreciate why using the standard ordinary least squares approach with nonlinear data will give you a meaningless result.

There is a pattern, but it cannot be easily described with a single straight line. The first few golf lessons appear to bring my score down rapidly. There is a negative association between lessons and my scores for this stretch; the slope is negative. More lessons yield lower scores (which is good in golf).

But then when I reach the point where I'm spending between $200 and $300 a month on lessons, the lessons do not seem to have much effect at all. There is no clear association over this stretch between additional instruction and my golf scores; the slope is zero.

And finally, the lessons appear to become counterproductive. Once I'm spending $300 a month on instruction, incremental lessons are associated with higher scores; the slope is positive over this stretch. (I'll discuss the distinct possibility that the bad golf may be causing the lessons, rather than the other way around, later in the chapter.)

The most important point here is that we cannot accurately summarize the relationship between lessons and scores with a single coefficient. The best interpretation of the pattern described above is that golf lessons have several different linear relationships with my scores. You can see that; a statistics package will not. If you feed these data into a regression equation, the computer will give you a single coefficient. That coefficient will not accurately reflect the true relationship between the variables of interest. The results you get will be the statistical equivalent of using a hair dryer in the bath tub.

Regression analysis is meant to be used when the relationship between variables is linear. A textbook or an advanced course in statistics will walk you through the other core assumptions underlying regression analysis. As with any other tool, the further one deviates from its intended use, the less effective, or even potentially dangerous, it's going to be.

Correlation does not equal causation. Regression analysis can only demonstrate an association between two variables. As I have mentioned before, we cannot prove with statistics alone that a change in one variable is *causing* a change in the other. In fact, a sloppy regression equation can produce a large and statistically significant association between two variables that have nothing to do with one another. Suppose we were searching for potential causes for the rising rate of autism in the United States over the last two decades. Our dependent variable—the outcome we are seeking to explain—would be some measure of the incidence of the autism by year, such as the number of diagnosed cases for every 1,000 children of a certain age. If we were to include annual per capita income in China as an explanatory variable, we would almost certainly find a positive and statistically significant association between rising incomes in China and rising autism rates in the U.S. over the past twenty years.

Why? Because they both have been rising sharply over the same period. Yet I highly doubt that a sharp recession in China would reduce the autism rate in the United States. (To be fair, if I observed a strong relationship between rapid economic growth in China and autism rates *in China alone*, I might begin to search for some environmental factor related to economic growth, such as industrial pollution, that might explain the association.)

The kind of false association between two variables that I have just illustrated is just one example of a more general phenomenon known as spurious causation. There are several other ways in which an association between A and B can be wrongly interpreted.

Reverse causality. A statistical association between A and B does not prove that A causes B. In fact, it's entirely plausible that B is causing A. I alluded to this possibility earlier in the golf lesson example. Suppose that when I build a complex model to explain my golf scores, the variable for golf lessons is consistently associated with worse scores. The more lessons I take, the worse I shoot! One explanation is that I have a really, really bad golf instructor. A more plausible explanation is that I tend to take more lessons when I'm playing poorly; bad golf is causing more lessons, not the other way around. (There are some simple methodological fixes to a problem of this nature. For example, I might include golf lessons in one month as an explanatory variable for golf scores *in the next month*.)

As noted earlier in the chapter, causality may go in both directions. Suppose you do some research demonstrating that states that spend more money on K–12 education have higher rates of economic growth than states that spend less on K–12 education. A positive and significant association between these two variables does not provide any insight into which direction the relationship happens to run. Investments in K–12 education could cause economic growth. On the other hand, states that have strong economies can afford to spend more on K–12 education, so the strong economy could be causing the education spending. Or, education spending could boost economic growth, which makes possible additional education spending—the causality could be going in both ways.

The point is that we should not use explanatory variables that might be affected by the outcome that we are trying to explain, or else the results will become hopelessly tangled. For example, it would be inappropriate to use the unemployment rate in a regression equation explaining GDP growth, since unemployment is clearly affected by the rate of GDP growth. Or, to think of it another way, a regression analysis finding that lowering unemployment will boost GDP growth is a silly and meaningless finding, since boosting GDP growth is usually required in order to reduce unemployment.

We should have reason to believe that our explanatory variables affect the dependent variable, and not the other way around.

Omitted variable bias. You should be skeptical the next time you see a huge headline proclaiming, "Golfers More Prone to Heart Disease, Cancer, and Arthritis!" I would not be surprised if golfers have a higher incidence of all of those diseases than nongolfers; I also suspect that golf is probably good for your health because it provides socialization and modest exercise. How can I reconcile those two statements? Very easily. Any study that attempts to measure the effects of playing golf on health must control properly for age. In general, people play more golf when they get older, particularly in retirement. Any analysis that leaves out age as an explanatory variable is going to miss the fact that golfers, on average, will be older than nongolfers. Golf isn't killing people; old age is killing people, and they happen to enjoy playing golf while it does. I suspect that when age is inserted into the regression analysis as a control variable, we will get a different outcome. *Among people who are the same age*, golf may be mildly preventive of serious illnesses. That's a pretty big difference.

In this example, age is an important "omitted variable." When we leave age out of a regression equation explaining heart disease or some other adverse health outcome, the "playing golf" variable takes on *two explanatory roles* rather than just one. It tells us the effect of playing golf on heart disease, and it tells us the effect of being old on heart disease (since golfers tend to be older than the rest of the population). In the statistics lingo, we would say that the golf variable is "picking up" the effect of age. The problem is that these two effects are comingled. At best, our results are a jumbled mess. At worst, we wrongly assume that golf is bad for your health, when in fact the opposite is likely to be true.

Regression results will be misleading and inaccurate if the regression equation leaves out an important explanatory variable, particularly if other variables in the equation "pick up" that effect. Suppose we are trying to explain school quality. This is an important outcome to understand: What makes good schools? Our dependent variable—the quantifiable measure of quality—would most likely be test scores. We would almost certainly examine school spending as one explanatory variable in hopes of quantifying the relationship between spending and test scores. Do schools that spend more get better results? If school spending were the only explanatory variable, I have no doubt that we would find a large and statistically significant relationship between spending and test scores. Yet that finding, and the implication that we can spend our way to better schools, is deeply flawed.

There are many potentially significant omitted variables here, but the crucial one is parental education. Well-educated families tend to live in affluent areas that spend a lot of money on their schools; such families also tend to have children who score well on tests (and poor families are more likely to have students who struggle). If we do not have some measure of the socioeconomic status of the student body as a control variable, our regression results will probably show a large positive association between school spending and test scores—when in fact, those results may be a function of the kind of students who are walking in the school door, not the money that is being spent in the building.

I remember a college professor's pointing out that SAT scores are highly correlated with the number of cars that a family owns. He insinuated that the SAT was therefore an unfair and inappropriate tool for college admissions. The SAT has its flaws but the correlation between scores and family cars is not the one that concerns me most. I do not worry much that rich families can get their kids into college by purchasing three extra automobiles. The number of cars in a family's garage is a proxy for their income, education, and other measures of socioeconomic status. The fact that wealthy kids do better on the SAT than poor kids is not news. (As noted earlier, the mean SAT critical reading score for students from families with a household income over $200,000 is 134 points higher than the mean score for students in households with income below $20,000.)[4] The bigger concern should be whether or not the SAT is "coachable." How much can students improve their scores by taking private SAT prep classes? Wealthy families clearly are better able to send their children to test prep classes. Any causal improvement between these classes and SAT scores would favor students from wealthy families relative to more disadvantaged students *of equal abilities* (who presumably also could have raised their scores with a prep class but never had that opportunity).

Highly correlated explanatory variables (multicollinearity). If a regression equation includes two or more explanatory variables that are highly correlated with each other, the analysis will not necessarily be able to discern the true relationship between each of those variables and the outcome that we are trying to explain. An example will make this clearer. Assume we are trying to gauge the effect of illegal drug use on SAT scores. Specifically, we have data on whether the participants in our study have ever used cocaine and also on whether they have ever used heroin. (We would presumably have many other control variables as well.) What is the impact of cocaine use on SAT scores, holding other factors constant, including heroin use? And what is the impact of heroin use on SAT scores, controlling for cocaine use and other factors?

The coefficients on heroin and cocaine use might not actually be able to tell us that. The methodological challenge is that people who have used heroin have likely also used cocaine. If we put both variables in the equation, we will have very few individuals who have used one drug but not the other, which leaves us very little variation in the data with which to calculate their independent effects. Think back for a moment to the mental imagery used to explain regression analysis in the last chapter. We divide our data sample into different "rooms" in which each observation is identical except for one variable, which then allows us to isolate the effect of that variable while controlling for other potential confounding factors. We may have 692 individuals in our sample who have used both cocaine and heroin. However, we may have only 3 individuals who have used cocaine but not heroin and 2 individuals who have used heroin and not cocaine. Any inference about the independent effect of just one drug or the other is going to be based on these tiny samples.

We are unlikely to get meaningful coefficients on either the cocaine or the heroin variable; we may also obscure the larger and more important relationship between SAT scores and using either one of these drugs. When two explanatory variables are highly correlated, researchers will usually use one or the other in the regression equation, or they may create some kind of composite variable, such as "used cocaine or heroin." For example, when researchers want to control for a student's overall socioeconomic background, they may include variables for both "mother's education" and "father's education," since this inclusion provides important insight into the educational background of the household. However, if the goal of the regression analysis is to isolate the effect of either a mother's or a father's education, then putting both variables into the equation is more likely to confuse the issue than to clarify it. The correlation between a husband's and a wife's educational attainments is so high that we cannot depend on regression analysis to give us coefficients that meaningfully isolate the effect of either parent's education (just as it is hard to separate the impact of cocaine use from the impact of heroin use).

Extrapolating beyond the data. Regression analysis, like all forms of statistical inference, is designed to offer us insights into the world around us. We seek patterns that will hold true for the larger population. *However, our results are valid only for a population that is similar to the sample on which the analysis has been done*. In the last chapter, I created a regression equation to predict weight based on a number of independent variables. The R^2 of my final model was .29, which means that it did a decent job of explaining the variation in weight for a large sample of individuals—all of whom happened to be adults.

So what happens if we use our regression equation to predict the likely weight of a newborn? Let's try it. My daughter was 21 inches when she was born. We'll say that her age at birth was zero; she had no education and did not exercise. She was white and female. The regression equation based on the Changing Lives data predicts that her weight at birth should have been negative 19.6 pounds. (She weighed 8½ pounds.)

The authors of one of the Whitehall studies referred to in the last chapter were strikingly explicit in drawing their narrow conclusion: "Low control in the work environment is associated with an increased risk of future coronary heart disease *among men and women employed in government offices*"[5] (italics added).

Data mining (too many variables). If omitting important variables is a potential problem, then presumably adding as many explanatory variables as possible to a regression equation must be the solution. Nope.

Your results can be compromised if you include too many variables, particularly extraneous explanatory variables with no theoretical justification. For example, one *should not* design a research strategy built around the following premise: Since we don't know what causes autism, we should put as many potential explanatory variables as possible in the regression equation just to see what might turn up as statistically significant; then maybe we'll get some answers. If you put enough junk variables in a regression equation, one of them is bound to meet the threshold for statistical significance just by chance. The further danger is that junk variables are not always easily recognized as such. Clever researchers can always build a theory *after the fact* for why some curious variable that is really just nonsense turns up as statistically significant.

To make this point, I often do the same coin flipping exercise that I explained during the probability discussion. In a class of forty students or so, I'll have each student flip a coin. Any student who flips tails is eliminated; the rest flip again. In the second round, those who flip tails are once again eliminated. I continue the rounds of flipping until one student has flipped five or six heads in a row. You may recall some of the silly follow-up questions: "What's your secret? Is it in the wrist? Can you teach us to flip heads all the time? Maybe it's that Harvard sweatshirt you're wearing."

Obviously the string of heads is just luck; the students have all watched it happen. However, that is not necessarily how the result could or would be interpreted in a scientific context. The probability of flipping five heads in a row is 1/32, or .03. This is comfortably below the .05 threshold we typically use to reject a null hypothesis. Our null hypothesis in this case is that the student has no special talent for flipping heads; the lucky string of heads (which is bound to happen for at least one student when I start with a large group) allows us to reject the null hypothesis and adopt the alternative hypothesis: This student has a special ability to flip heads. After he has achieved this impressive feat, we can study him for clues about his flipping success—his flipping form, his athletic training, his extraordinary concentration while the coin is in the air, and so on. And it is all nonsense.

This phenomenon can plague even legitimate research. The accepted convention is to reject a null hypothesis when we observe something that would happen by chance only 1 in 20 times or less if the null hypothesis were true. Of course, if we conduct 20 studies, or if we include 20 junk variables in a single regression equation, then on average we will get 1 bogus statistically significant finding. The *New York Times Magazine* captured this tension wonderfully in a quotation from Richard Peto, a medical statistician and epidemiologist: "Epidemiology is so beautiful and provides such an important perspective on human life and death, but an incredible amount of rubbish is published."[6]

Even the results of clinical trials, which are usually randomized experiments and therefore the gold standard of medical research, should be viewed with some skepticism. In 2011, the *Wall Street Journal* ran a front-page story on what it described as one of the "dirty little secrets" of medical research: "Most results, including those that appear in top-flight peer-reviewed journals, can't be reproduced."[7] (A peer-reviewed journal is a publication in which studies and articles are reviewed for methodological soundness by other experts in the same field before being approved for publication; such publications are considered to be the gatekeepers for academic research.) One reason for this "dirty little secret" is the positive publication bias described in Chapter 7. If researchers and medical journals pay attention to positive findings and ignore negative findings, then they may well publish the one study that finds a drug effective and ignore the nineteen in which it has no effect. Some clinical trials may also have small samples (such as for a rare diseases), which magnifies the chances that random variation in the data will get more attention than it deserves. On top of that, researchers may have some conscious or unconscious bias, either because of a strongly held prior belief or because a positive finding would be better for their career. (No one ever gets rich or famous by proving what *doesn't* cure cancer.)

For all of these reasons, a shocking amount of expert research turns out to be wrong. John Ioannidis, a Greek doctor and epidemiologist, examined forty-nine studies published in three prominent medical journals.[8] Each study had been cited in the medical literature at least a thousand times. Yet roughly one-third of the research was subsequently refuted by later work. (For example, some of the studies he examined promoted estrogen replacement therapy.) Dr. Ioannidis estimates that roughly half of the scientific papers published will eventually turn out to be wrong.[9] His research was published in the *Journal of the American Medical Association*, one of the journals in which the articles he studied had appeared. This does create a certain mind-bending irony: If Dr. Ioannidis's research is correct, then there is a good chance that his research is wrong.

Regression analysis is still an awesome statistical tool. (Okay, perhaps my description of it as a "miracle elixir" in the last chapter was a little hyperbolic.) Regression analysis enables us to find key patterns in large data sets, and those patterns are often the key to important research in medicine and the social sciences. Statistics gives us objective standards for evaluating these patterns. When used properly, regression analysis is an important part of the scientific method. Consider this chapter to be the mandatory warning label.

All of the assorted specific warnings on that label can be boiled down to two key lessons. First, designing a good regression equation—figuring out what variables should be examined and where the data should come from—is more important than the underlying statistical calculations. This process is referred to as estimating the equation, or specifying a good regression equation. The best researchers are the ones who can think logically about what variables ought to be included in a regression equation, what might be missing, and how the eventual results can and should be interpreted.

Second, like most other statistical inference, regression analysis builds only a circumstantial case. An association between two variables is like a fingerprint at the scene of the crime. It points us in the right direction, but it's rarely enough to convict. (And sometimes a fingerprint at the scene of a crime *doesn't* belong to the perpetrator.) Any regression analysis needs a theoretical underpinning: Why are the explanatory variables in the equation? What phenomena from other disciplines can explain the observed results? For instance, why do we think that wearing purple shoes would boost performance on the math portion of the SAT or that eating popcorn can help prevent prostate cancer? The results need to be replicated, or at least consistent with other findings.

Even a miracle elixir won't work when not taken as directed.

CHAPTER 13

Program Evaluation

Will going to Harvard change your life?

Brilliant researchers in the social sciences are not brilliant because they can do complex calculations in their heads, or because they win more money on *Jeopardy* than less brilliant researchers do (though both these feats may be true). Brilliant researchers—those who appreciably change our knowledge of the world—are often individuals or teams who find creative ways to do "controlled" experiments. To measure the effect of any treatment or intervention, we need something to measure it against. How would going to Harvard affect your life? Well, to answer that question, we have to know what happens to you after you go to Harvard—and what happens to you *after you don't go to Harvard*. Obviously we can't have data on both. Yet clever researchers find ways to compare some treatment (e.g., going to Harvard) with the counterfactual, which is what would have happened in the absence of that treatment.

To illustrate this point, let's ponder a seemingly simple question: Does putting more police officers on the street deter crime? This is a socially significant question, as crime imposes huge costs on society. If a greater police presence lowers crime, either through deterrence or by catching and imprisoning bad guys, then investments in additional police officers could have large returns. On the other hand, police officers are relatively expensive; if they have little or no impact on crime reduction, then society could make better use of its resources elsewhere (perhaps with investments in crime-fighting technology, such as surveillance cameras).

The challenge is that our seemingly simple question—what is the causal effect of more police officers on crime?—turns out to be very difficult to answer. By this point in the book, you should recognize that we cannot answer this question simply by examining whether jurisdictions with a high number of police officers per capita have lower rates of crime. Zurich is not Los Angeles. Even a comparison of large American cities will be deeply flawed; Los Angeles, New York, Houston, Miami, Detroit, and Chicago are all different places with different demographics and crime challenges.

Our usual approach would be to attempt to specify a regression equation that controls for these differences. Alas, even multiple regression analysis is not going to save us here. If we attempt to explain crime rates (our dependent variable) by using police officers per capita as an explanatory variable (along with other controls), we will have a serious reverse causality problem. We have a solid theoretical reason to believe that putting more police officers on the street will reduce crime, but it's also possible that crime could "cause" police officers, in the sense that cities experiencing crime waves will hire more police officers. We could easily find a positive but misleading association between crime and police: the places with the most police officers have the worst crime problems. Of course, the places with lots of doctors also tend to have the highest concentration of sick people. These doctors aren't making people sick; they are located in places where they are needed most (and at the same time sick people are moving to places where they can get appropriate medical care). I suspect that there are disproportionate numbers of oncologists and cardiologists in Florida; banishing them from the state will not make the retiree population healthier.

Welcome to program evaluation, which is the process by which we seek to measure the causal effect of some intervention—anything from a new cancer drug to a job placement program for high school dropouts. Or putting more police officers on the street. The intervention that we care about is typically called the "treatment," though that word is used more expansively in a statistical context than in normal parlance. A treatment can be a literal treatment, as in some kind of medical intervention, or it can be something like attending college or receiving job training upon release from prison. The point is that we are seeking to isolate the effect of that single factor; ideally we would like to know how the group receiving that treatment fares compared with some other group whose members are identical in all other respects but for the treatment.

Program evaluation offers a set of tools for isolating the treatment effect when cause and effect are otherwise elusive. Here is how Jonathan Klick and Alexander Tabarrok, researchers at the University of Pennsylvania and George Mason University, respectively, studied how putting more police officers on the street affects the crime rate. Their research strategy made use of the terrorism alert system. Specifically, Washington, D.C., responds to "high alert" days for terrorism by putting more officers in certain areas of the city, since the capital is a natural terrorism target. We can assume that there is no relationship between street crime and the terrorism threat, so this boost in the D.C. police presence is unrelated to the conventional crime rate, or "exogenous." The researchers' most valuable insight was recognizing the natural experiment here: What happens to *ordinary crime* on terrorism "high alert" days?

The answer: The number of crimes committed when the terrorism threat was Orange (high alert and more police) was roughly 7 percent lower than when the terrorism threat level was Yellow (elevated alert but no extra law enforcement precautions). The authors also found that the decrease in crime was sharpest in the police district that gets the most police attention on high-alert days (because it includes the White House, the Capitol, and the National Mall). The important takeaway is that we can answer tricky but socially meaningful questions—we just have to be clever about it. Here are some of the most common approaches for isolating a treatment effect.

Randomized, controlled experiments. The most straightforward way to create a treatment and control group is to—wait for it—create a treatment and control group. There are two big challenges to this approach. First, there are many kinds of experiments that we cannot perform on people. This constraint (I hope) is not going away anytime soon. As a result, we can do controlled experiments on human subjects only when there is reason to believe that the treatment effect has a potentially positive outcome. This is often *not* the case (e.g., "treatments" like experimenting with drugs or dropping out of high school), which is why we need the strategies introduced in the balance of the chapter.

Second, there is a lot more variation among people than among laboratory rats. The treatment effect that we are testing could easily be confounded by other variations in the treatment and control groups; you are bound to have tall people, short people, sick people, healthy people, males, females, criminals, alcoholics, investment bankers, and so on. How can we ensure that differences across these other characteristics don't mess up the results? I have good news: This is one of those rare instances in life in which the best approach involves the least work! The optimal way to create any treatment and control group is to distribute the study participants randomly across the two groups. The beauty of randomization is that it will generally distribute the non-treatment-related variables more or less evenly between the two groups—both the characteristics that are obvious, such as sex, race, age, and education and the nonobservable characteristics that might otherwise mess up the results.

Think about it: If we have 1,000 females in our prospective sample, then when we split the sample randomly into two groups, the most likely outcome is that 500 females will end up in each. Obviously we can't expect that split exactly, but once again probability is our friend. The probability is low that one group will get a disproportionate number of women (or a disproportionate number of individuals with any other characteristic). For example, if we have a sample of 1,000 people, half of whom are women, there is less than a 1 percent chance of getting fewer than 450 women in one group or the other. Obviously the bigger the samples, the more effective randomization will be in creating two broadly similar groups.

Medical trials typically aspire to do randomized, controlled experiments. Ideally these clinical trials are double-blind, meaning that neither the patient nor the physician knows who is receiving the treatment and who is getting a placebo. This is obviously impossible with treatments such as surgical procedures (the heart surgeon will, one hopes, know which patients are getting bypass surgery). Even with surgical procedures, however, it may still be possible to keep patients from learning whether they are in the treatment or the control group. One of my favorite studies involved an evaluation of a certain kind of knee surgery to alleviate pain. The treatment group was given the surgery. The control group was given a "sham" surgery in which the surgeon made three small incisions in the knee and "pretended to operate."* It turned out that the real surgery was no more effective than the sham surgery in relieving knee pain.[1]

Randomized trials can be used to test some interesting phenomena. For example, do prayers offered by strangers improve postsurgical outcomes? Reasonable people have widely varying views on religion, but a study published in the *American Heart Journal* conducted a controlled study that examined whether patients recovering from heart bypass surgery would have fewer postoperative complications if a large group of strangers prayed for their safe and speedy recovery.[2] The study involved 1,800 patients and members of three religious congregations from across the country. The patients, all of whom received coronary bypass surgery, were divided into three groups: one group was not prayed for; one group was prayed for and was told so; the third group was prayed for, but the participants in that group were told that they might or might not receive prayers (thereby controlling for a prayer placebo effect). Meanwhile, the members of the religious congregations were told to offer prayers for specific patients by first name and the first initial of their last name (e.g., Charlie W.). The congregants were given latitude in how they prayed, so long as the prayer included the phrase "for a successful surgery with a quick, healthy recovery and no complications."

And? Will prayer be the cost-effective solution to America's health care challenges? Probably not. The researchers did not find any difference in the rate of complications within thirty days of surgery for those who were offered prayers compared with those who were not. Critics of the study pointed out a potential omitted variable: prayers coming from other sources. As the *New York Times* summarized, "Experts said the

---

* The participants did know that they were participating in a clinical trial and might receive the sham surgery.

study could not overcome perhaps the largest obstacle to prayer study: the unknown amount of prayer each person received from friends, families, and congregations around the world who pray daily for the sick and dying."

Experimenting on humans can get you arrested, or perhaps hauled off to appear before some international criminal tribunal. You should be aware of this. However, there is still room in the social sciences for randomized, controlled experiments involving "human subjects." One famous and influential experiment is the Tennessee Project STAR experiment, which tested the effect of smaller class sizes on student learning. The relationship between class size and learning is hugely important. Nations around the world are struggling to improve educational outcomes. If smaller classes promote more effective learning, ceteris paribus, then society ought to invest in hiring more teachers to bring class sizes down. At the same time, hiring teachers is expensive; if students in smaller classes are doing better for reasons *unrelated to the size of the class*, then we could end up wasting an enormous amount of money.

The relationship between class size and student achievement is surprisingly hard to study. Schools with small classes generally have greater resources, meaning that both the students and the teachers are likely to be different from students and teachers in schools with larger classes. And within schools, smaller classes tend to be smaller for a reason. A principal may assign difficult students to a small class, in which case we might find a spurious negative association between smaller classes and student achievement. Or veteran teachers may choose to teach small classes, in which case the benefit of small classes may come from the teachers who choose to teach them rather than from the lower pupil-teacher ratio.

Beginning in 1985, Tennessee's Project STAR did a controlled experiment to test the effects of smaller classes.[3] (Lamar Alexander was governor of Tennessee at the time; he later went on to become secretary of education under President George H. W. Bush.) In kindergarten, students in seventy-nine different schools were randomly assigned to either a small class (13–17 students), a regular class (22–25 students), or a regular class with both a regular teacher and a teacher's aide. Teachers were also randomly assigned to the different classrooms. Students stayed in the class type to which they were randomly assigned through third grade. Assorted life realities chipped away at the randomization. Some students entered the system in the middle of the experiment; others left. Some students were moved from class to class for disciplinary reasons; some parents lobbied successfully to have students moved to smaller classes. And so on.

Still, Project STAR remains the only randomized test of the effects of smaller classes. The results turned out to be statistically and socially significant. Overall, students in the small classes performed .15 standard deviations better on standardized tests than students in the regular-size classes; black students in small classes had gains that were twice as large. Now the bad news. The Project STAR experiment cost roughly $12 million. The study on the effect of prayer on postsurgical complications cost $2.4 million. The finest studies are like the finest of anything else: They cost big bucks.

Natural experiment. Not everybody has millions of dollars lying around to create a large, randomized trial. A more economical alternative is to exploit a natural experiment, which happens when random circumstances somehow create something approximating a randomized, controlled experiment. This was the case with our Washington, D.C., police example at the beginning of the chapter. Life sometimes creates a treatment and control group by accident; when that occurs, researchers are eager to leap on the results. Consider the striking but complicated link between education and longevity. People who get more education tend to live longer, even after controlling for things like income and access to health care. As the *New York Times* has noted, "The one social factor that researchers agree is consistently linked to longer lives in every country where it has been studied is education. It is more important than race; it obliterates any effects of income."[4] But so far, that's just a correlation. Does more education, ceteris paribus, cause better health? If you think of the education itself as the "treatment," will getting more education make you live longer?

This would appear to be a nearly impossible question to study, since people who choose to get more education are different from people who don't. The difference between high school graduates and college graduates is not just four years of schooling. There could easily be some unobservable characteristics shared by people who pursue education that also explain their longer life expectancy. If that is the case, offering more education to those who would have chosen less education won't actually improve their health. The improved health would not be a function of the incremental education; it would be a function of the kind of people who pursue that incremental education.

We cannot conduct a randomized experiment to solve this conundrum, because that would involve making some participants leave school earlier than they would like. (You try explaining to someone that he can't go to college—ever—because he is in the control group.) The only possible test of the causal effect of education on longevity would be some kind of experiment that forced a large segment of the population to stay in school longer than its members might otherwise choose. That's at least morally acceptable since we expect a positive treatment effect. Still, we can't force kids to stay in school; that's not the American way.

Oh, but it is. Every state has some kind of minimum schooling law, and at different points in history *those laws have changed.* That kind of exogenous change in schooling attainment—meaning that it's not caused by the individuals being studied—is exactly the kind of thing that makes researchers swoon with excitement. Adriana Lleras-Muney, a graduate student at Columbia, saw the research potential in the fact that different states have changed their minimum schooling laws at different points in time. She went back in history and studied the relationship between when states changed their minimum schooling laws and later changes in life expectancy in those states (by trolling through lots and lots of census data). She still had a methodological challenge; if the residents of a state live longer after the state raises its minimum schooling law, we cannot attribute the longevity to the extra schooling. Life expectancy is generally going up over time. People lived longer in 1900 than in 1850, no matter what the states did.

However, Lleras-Muney had a natural control: states that did *not* change their minimum schooling laws. Her work approximates a giant laboratory experiment in which the residents of Illinois are forced to stay in school for seven years while their neighbors in Indiana can leave school after six years. The difference is that this controlled experiment was made possible by a historical accident—hence the term "natural experiment."

What happened? Life expectancy of those adults who reached age thirty-five was extended by an extra one and a half years just by their attending one additional year of school.[5] Lleras-Muney's results have been replicated in other countries where variations in mandatory schooling laws have created similar natural experiments. Some skepticism is in order. We still do not understand the mechanism by which additional schooling leads to longer lives.

Nonequivalent control. Sometimes the best available option for studying a treatment effect is to create nonrandomized treatment and control groups. Our hope/expectation is that the two groups are broadly similar even though circumstances have not allowed us the statistical luxury of randomizing. The good news is that we have a treatment and a control group. The bad news is that any nonrandom assignment creates at least the potential for bias. There may be unobserved differences between the treatment and control groups related to how participants are assigned to one group or the other. Hence the name "nonequivalent control."

A nonequivalent control group can still be a very helpful tool. Let's ponder the question posed in the title of this chapter: Is there a significant life advantage to attending a highly selective college or university? Obviously the Harvard, Princeton, and Dartmouth graduates of the world do very well. On average, they earn more money and have more expansive life opportunities than students who attend less selective institutions. (A 2008 study by PayScale.com found that the median pay for Dartmouth graduates with ten to twenty years of work experience was $134,000, the highest of any undergraduate institution; Princeton was second with a median of $131,000.)[6] As I hope you realize by this point, these impressive numbers tell us absolutely nothing about the value of a Dartmouth or Princeton education. Students who attend Dartmouth and Princeton are talented when they apply; *that's why they get accepted*. They would probably do well in life no matter where they went to college.

What we don't know is the treatment effect of attending a place like Harvard or Yale. Do the graduates of these elite institutions do well in life because they were hyper-talented when they walked onto the campus? Or do these colleges and universities add value by taking talented individuals and making them even more productive? Or both?

We cannot conduct a randomized experiment to answer this question. Few high school students would agree to be randomly assigned to a college; nor would Harvard and Dartmouth be particularly keen about taking the students randomly assigned to them. We appear to be left without any mechanism for testing the value of the treatment effect. Cleverness to the rescue! Economists Stacy Dale and Alan Krueger found a way to answer this question by exploiting* the fact that many students apply to multiple colleges.[7] Some of those students are accepted at a highly selective school and choose to attend that school; others are accepted at a highly selective school but choose to attend a less selective college or university instead. Bingo! Now we have a treatment group (those students who attended highly selective colleges and universities) and a nonequivalent control group (those students *who were talented enough to be accepted by such a school* but opted to attend a less selective institution instead).†

---

* Researchers love to use the word "exploit." It has a specific meaning in terms of taking advantage of some data-related opportunity. For example, when researchers find some natural experiment that creates a treatment and control group, they will describe how they plan to "exploit the variation in the data."

† There is potential for bias here. Both groups of students are talented enough to get into a highly selective school. However, one group of students chose to go to such a school, and the other group did not. The group of students who chose to attend a less selective school may be less motivated, less hardworking, or different in some other ways that we cannot observe. If Dale and Krueger had found that students who attend a highly selective school had higher lifetime earnings than students who were accepted at such a school but went to a less selective college instead, we still could not be certain whether the difference was due to the selective school or to the kind of student who opted to attend such a school when given a choice. This potential bias turns out to be unimportant in the Dale and Krueger study, however, because of its direction. Dale and Krueger find that the students who attended highly selective schools did not earn significantly more in life than students who were accepted but went elsewhere *despite the fact that the students who declined to attend a highly selective*

Dale and Krueger studied longitudinal data on the earnings of both groups. This is not a perfect apples-and-apples comparison, and earnings are clearly not the only life outcome that matters, but their findings should assuage the anxieties of overwrought high school students and their parents. Students who attended more selective colleges earned roughly the same as students of seemingly similar ability who attended less selective schools. The one exception was students from low-income families, who earned more if they attended a selective college or university. The Dale and Krueger approach is an elegant way to sort out the treatment effect (spending four years at an elite institution) from the selection effect (the most talented students are admitted to those institutions). In a summary of the research for the *New York Times*, Alan Krueger indirectly answered the question posed in the title of this chapter, "Recognize that your own motivation, ambition, and talents will determine your success more than the college name on your diploma."[8]

Difference in differences. One of the best ways to observe cause and effect is to do something and then see what happens. This is, after all, how infants and toddlers (and sometimes adults) learn about the world. My children were very quick to learn that if they hurled pieces of food across the kitchen (cause), the dog would race eagerly after them (effect). Presumably the same power of observation can help inform the rest of life. If we cut taxes and the economy improves, then the tax cuts must have been responsible.

Maybe. The enormous potential pitfall with this approach is that life tends to be more complex than throwing chicken nuggets across the kitchen. Yes, we may have cut taxes at a specific point in time, but there were other "interventions" unfolding during roughly the same stretch: More women were going to college, the Internet and other technological innovations were raising the productivity of American workers, the Chinese currency was undervalued, the Chicago Cubs fired their general

---

school may have had attributes that caused them to earn less in life apart from their education. If anything, the bias here causes the findings to *overstate* the pecuniary benefits of attending a highly selective college—which turn out to be insubstantial anyway.

manager, and so on. Whatever happened after the tax cut cannot be attributed solely to the tax cut. The challenge with any "before and after" kind of analysis is that just because one thing follows another does not mean that there is a causal relationship between the two.

A "difference in differences" approach can help us identify the effects of some intervention by doing two things. First, we examine the "before" and "after" data for whatever group or jurisdiction has received the treatment, such as the unemployment figures for a county that has implemented a job training program. Second, we compare those data with the unemployment figures over the same time period for a similar county that did not implement any such program.

The important assumption is that the two groups used for the analysis are largely comparable except for the treatment; as a result, any significant difference in outcomes between the two groups can be attributed to the program or policy being evaluated. For example, suppose that one county in Illinois implements a job training program to combat high unemployment. Over the ensuing two years, the unemployment rate continues to rise. Does that make the program a failure? Who knows?

Effect of Job Training on Unemployment in County A

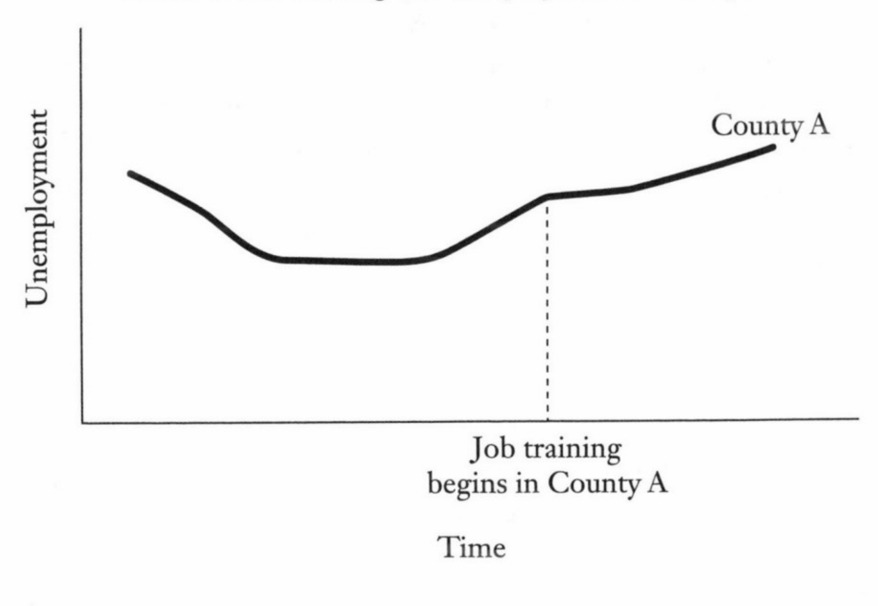

Other broad economic forces may be at work, including the possibility of a prolonged economic slump. A difference-in-differences approach would compare the change in the unemployment rate over time in the county we are evaluating with the unemployment rate for a neighboring county with no job training program; the two counties must be similar in all other important ways: industry mix, demographics, and so on. How does the unemployment rate in the county with the new job training program change over time *relative to the county that did not implement such a program*? We can reasonably infer the treatment effect of the program by comparing the changes in the two counties over the period of study—the "difference in differences." The other county in this study is effectively acting as a control group, which allows us to take advantage of the data collected before and after the intervention. If the control group is good, it will be exposed to the same broader forces as our treatment group. The difference-in-differences approach can be particularly enlightening when the treatment initially appears ineffective (unemployment is higher after the program is implemented than before), yet the control group shows us that the trend would have been even worse in the absence of the intervention.

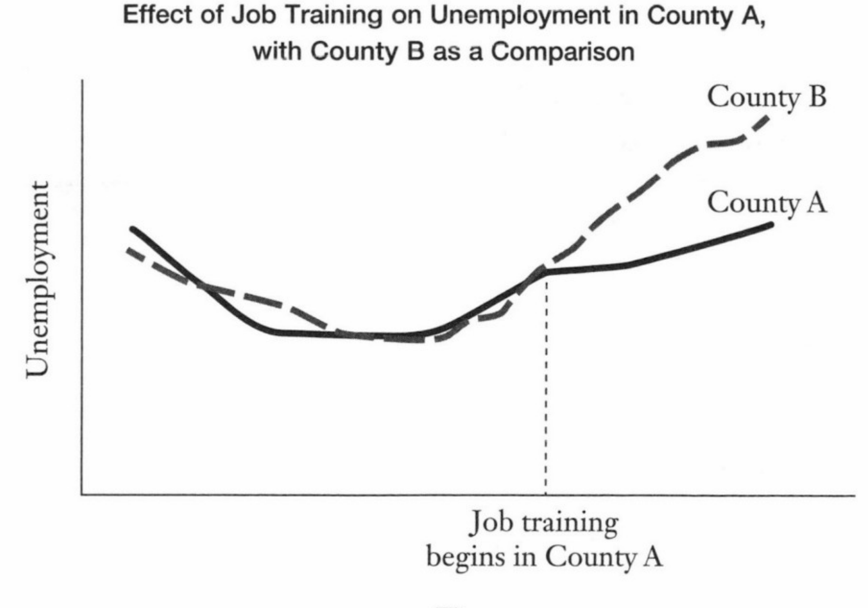

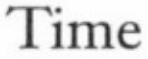

Discontinuity analysis. One way to create a treatment and control group is to compare the outcomes for some group that barely qualified for an intervention or treatment with the outcomes for a group that just missed the cutoff for eligibility and did not receive the treatment. Those individuals who fall just above and just below some arbitrary cutoff, such as an exam score or a minimum household income, will be nearly identical in many important respects; the fact that one group received the treatment and the other didn't is essentially arbitrary. As a result, we can compare their outcomes in ways that provide meaningful results about the effectiveness of the relevant intervention.

Suppose a school district requires summer school for struggling students. The district would like to know whether the summer program has any long-term academic value. As usual, a simple comparison between the students who attend summer school and those who do not would be worse than useless. The students who attend summer school are there *because they are struggling*. Even if the summer school program is highly effective, the participating students will probably still do worse in the long run than the students who were not required to take summer school. What we want to know is how the struggling students perform after taking summer school *compared with how they would have done if they had not taken summer school*. Yes, we could do some kind of controlled experiment in which struggling students are randomly selected to attend summer school or not, but that would involve denying the control group access to a program that we think would be helpful.

Instead, the treatment and control groups are created by comparing those students who just barely fell below the threshold for summer school with those who just barely escaped it. Think about it: the students who fail a midterm are appreciably different from students who do not fail the midterm. But students who get a 59 percent (a failing grade) are *not* appreciably different from those students who get a 60 percent (a passing grade). If those who fail the midterm are enrolled in some treatment, such as mandatory tutoring for the final exam, then we would have a reasonable treatment and control group if we compared the final exam scores of those who barely failed the midterm (and received tutoring) with the scores of those who barely passed the midterm (and did not get tutoring).

This approach was used to determine the effectiveness of incarceration for juvenile offenders as a deterrent to future crime. Obviously this kind of analysis cannot simply compare the recidivism rates of juvenile offenders who are imprisoned with the recidivism rates for juvenile offenders who received lighter sentences. *The juvenile offenders who are sent to prison typically commit more serious crimes than the juvenile offenders who receive lighter sentences; that's why they go to prison.* Nor can we create a treatment and control group by distributing prison sentences randomly (unless you want to risk twenty-five years in the big house the next time you make an illegal right turn on red). Randi Hjalmarsson, a researcher now at the University of London, exploited rigid sentencing guidelines for juvenile offenders in the state of Washington to gain insight into the causal effect of a prison sentence on future criminal behavior. Specifically, she compared the recidivism rate for those juvenile offenders who were "just barely" sentenced to prison with the recidivism rate for those juveniles who "just barely" got a pass (which usually involved a fine or probation).[9]

The Washington criminal justice system creates a grid for each convicted offender that is used to administer a sentence. The x-axis measures the offender's prior adjudicated offenses. For example, each prior felony counts as one point; each prior misdemeanor counts as one-quarter point. The point total is rounded down to a whole number (which will matter in a moment). Meanwhile, the y-axis measures the severity of the current offense on a scale from E (least serious) to A+ (most serious). A convicted juvenile's sentence is literally calculated by finding the appropriate box on the grid: An offender with two points' worth of prior offenses who commits a Class B felony will receive fifteen to thirty-six months in a juvenile jail. A convicted offender with only one point worth of prior offenses who commits the same crime will not be sent to jail. That discontinuity is what motivated the research strategy. Hjalmarsson compared the outcomes for convicted offenders who fell just above and below the threshold for a jail sentence. As she explains in the paper, "If there are two individuals with a current offense class of C+ and [prior] adjudication scores of 2¾ and 3, then only the latter individual will be sentenced to state incarceration."

For research purposes, those two individuals are essentially the same—until one of them goes to jail. And at that point, their behavior does appear to diverge sharply. The juvenile offenders who go to jail are significantly less likely to be convicted of another crime (after they are released from jail).

We care about what works. This is true in medicine, in economics, in business, in criminal justice—in everything. Yet causality is a tough nut to crack, even in cases where cause and effect seems stunningly obvious. To understand the true impact of a treatment, we need to know the "counterfactual," which is what would have happened in the absence of that treatment or intervention. Often the counterfactual is difficult or impossible to observe. Consider a nonstatistics example: Did the U.S. invasion of Iraq make America safer?

There is only one intellectually honest answer: We will never know. The reason we will never know is that we do not know—and cannot know—what would have happened if the United States had not invaded Iraq. True, the United States did not find weapons of mass destruction. But it is possible that on the day after the United States *did not invade Iraq* Saddam Hussein could have climbed into the shower and said to himself, "I could really use a hydrogen bomb. I wonder if the North Koreans will sell me one?" After that, who knows?

Of course, it's also possible that Saddam Hussein could have climbed into that same shower on the day after the United States *did not invade Iraq* and said to himself, "I could really use—" at which point he slipped on a bar of soap, hit his head on an ornate marble fixture, and died. In that case, the world would have been rid of Saddam Hussein without the enormous costs associated with the U.S. invasion. Who knows what would have happened?

The purpose of any program evaluation is to provide some kind of counterfactual against which a treatment or intervention can be measured. In the case of a randomized, controlled experiment, the control group is the counterfactual. In cases where a controlled experiment is impractical or immoral, we need to find some other way of approximating the counterfactual. Our understanding of the world depends on finding clever ways to do that.

Conclusion

Five questions that statistics can help answer

Not that long ago, information was much harder to gather and far more expensive to analyze. Imagine studying the information from one million credit card transactions in the era—only a few decades back—when there were merely paper receipts and no personal computers for analyzing the accumulated data. During the Great Depression, there were no official statistics with which to gauge the depth of the economic problems. Government did not collect official information on either gross domestic product (GDP) or unemployment, meaning that politicians were attempting to do the economic equivalent of navigating through a forest without a compass. Herbert Hoover declared that the Great Depression was over in 1930, on the basis of the inaccurate and outdated data that were available. He told the country in his State of Union address that two and a half million Americans were out of work. In fact, five million Americans were jobless, and unemployment was climbing by one hundred thousand every week. As James Surowiecki recently observed in *The New Yorker*, "Washington was making policy in the dark."[1]

We are now awash in data. For the most part, that is a good thing. The statistical tools introduced in this book can be used to address some of our most significant social challenges. In that vein, I thought it fitting to finish the book with questions, not answers. As we try to digest and analyze staggering quantities of information, here are five important (and admittedly random) questions whose socially significant answers will involve many of the tools introduced in this book.

WHAT IS THE FUTURE OF FOOTBALL?

In 2009, Malcolm Gladwell posed a question in a *New Yorker* article that first struck me as needlessly sensationalist and provocative: How different are dog fighting and football?[2] The connection between the two activities stemmed from the fact that quarterback Michael Vick, who had served time in prison for his involvement in a dog-fighting ring, had been reinstated in the National Football League just as information was beginning to emerge that football-related head trauma may be associated with depression, memory loss, dementia, and other neurological problems later in life. Gladwell's central premise was that both professional football and dog fighting are inherently devastating to the participants. By the end of the article, I was convinced that he had raised an intriguing point.

Here is what we know. There is mounting evidence that concussions and other brain injuries associated with playing football can cause serious and permanent neurological damage. (Similar phenomena have been observed in boxers and hockey players.) Many prominent former NFL players have shared publicly their post-football battles with depression, memory loss, and dementia. Perhaps the most poignant was Dave Duerson, a former safety and Super Bowl winner for the Chicago Bears, who committed suicide by shooting himself in the chest; he left explicit instructions for his family to have his brain studied after his death.

In a phone survey of a thousand randomly selected former NFL players who had played at least three years in the league, 6.1 percent of the former players over fifty reported that they had received a diagnosis of "dementia, Alzheimer's disease, or other memory-related disease." That's five times the national average for that age group. For younger players, the rate of diagnosis was nineteen times the national average. Hundreds of former NFL players have now sued both the league and the makers of football helmets for allegedly hiding information about the dangers of head trauma.[3]

One of the researchers studying the impacts of brain trauma is Ann McKee, who runs the neuropathology laboratory at the Veterans Hospital in Bedford, Massachusetts. (Coincidentally, McKee also does the neuropathology work for the Framingham Heart Study.) Dr. McKee has documented the buildup of abnormal proteins called tau in the brains of athletes who have suffered brain trauma, such as boxers and football players. This leads to a condition known as chronic traumatic encephalopathy, or CTE, which is a progressive neurological disorder that has many of the same manifestations as Alzheimer's.

Meanwhile, other researchers have been documenting the connection between football and brain trauma. Kevin Guskiewicz, who runs the Sports Concussion Research Program at the University of North Carolina, has installed sensors on the inside of the helmets of North Carolina football players to record the force and nature of blows to the head. According to his data, players routinely receive blows to the head with a force equivalent to hitting the windshield in a car crash at twenty-five miles per hour.

Here is what we don't know. Is the brain injury evidence uncovered so far representative of the long-term neurological risks that all professional football players face? Or might this just be a "cluster" of adverse outcomes that is a statistical aberration? Even if it turns out that football players do face significantly higher risks of neurological disorder later in life, we would still have to probe the causality. Might the kind of men who play football (and boxing and hockey) be prone to this kind of problem? Is it possible that some other factors, such as steroid use, are contributing to the neurological problems later in life?

If the accumulating evidence does suggest a clear, causal link between playing football and long-term brain injury, one overriding question will have to be addressed by players (and the parents of younger players), coaches, lawyers, NFL officials, and perhaps even government regulators: Is there a way to play the game of football that reduces most or all of the head trauma risk? If not, then what? This is the point behind Malcolm Gladwell's comparison of football and dog fighting. He explains that dog fighting is abhorrent to the public because the dog owner willingly submits his dog to a contest that culminates in suffering and destruction. "And why?" he asks. "For the entertainment of an audience and the chance of a payday. In the nineteenth century, dog fighting was widely accepted by the American public. But we no longer find that kind of transaction morally acceptable in a sport."

Nearly every kind of statistical analysis described in this book is currently being used to figure out whether or not professional football as we know it now has a future.

WHAT (IF ANYTHING) IS CAUSING THE DRAMATIC RISE IN THE INCIDENCE OF AUTISM?

In 2012, the Centers for Disease Control reported that 1 in 88 American children has been diagnosed with an autism spectrum disorder (on the basis of data from 2008).[4] The rate of diagnosis had climbed from 1 in 110 in 2006, and 1 in 150 in 2002—or nearly a doubling in less than a decade. Autism spectrum disorders (ASDs) are a group of developmental disabilities characterized by atypical development in socialization, communication, and behavior. The "spectrum" indicates that autism encompasses a broad range of behaviorally defined conditions.[5] Boys are five times as likely to be diagnosed with an ASD as girls (meaning that the incidence for boys is even higher than 1 in 88).

The first intriguing statistical question is whether we are experiencing an epidemic of autism, an "epidemic of diagnosis," or some combination of the two?[6] In previous decades, children with an autism spectrum disorder had symptoms that might have gone undiagnosed, or their developmental challenges might have been described more generally as a "learning disability." Doctors, parents, and teachers are now much more aware of the symptoms of ASDs, which naturally leads to more diagnoses regardless of whether or not the incidence of autism is on the rise.

In any case, the shockingly high incidence of ASDs represents a serious challenge for families, for schools, and for the rest of society. The average lifetime cost of managing an autism spectrum disorder for a single individual is $3.5 million.[7] Despite what is clearly an epidemic, we know amazingly little about what causes the condition. Thomas Insel, director of the National Institute of Mental Health, has said, "Is it cell phones? Ultrasound? Diet sodas? Every parent has a theory. At this point, we just don't know."[8]

What is different or unique about the lives and backgrounds of children with ASDs? What are the most significant physiological differences between children with and without an ASD? Is the incidence of ASDs different across countries? If so, why? Traditional statistical detective work is turning up clues.

One recent study by researchers at the University of California at Davis identified ten locations in California with autism rates that are double the rates of surrounding areas; each of the autism clusters is a neighborhood with a concentration of white, highly educated parents.[9] Is that a clue, or a coincidence? Or might it reflect that relatively privileged families are more likely to have an autism spectrum disorder diagnosed? The same researchers are also conducting a study in which they will collect dust samples from the homes of 1,300 families with an autistic child to test for chemicals or other environmental contaminants than may play a causal role.

Meanwhile, other researchers have identified what appears to be a genetic component to autism by studying ASDs among identical and fraternal twins.[10] The likelihood that two children in the same family have an ASD is higher among identical twins (who share the same genetic makeup) than among fraternal twins (whose genetic similarity is the same as for regular siblings). This finding does not rule out significant environmental factors, or perhaps the interaction between environmental and genetic factors. After all, heart disease has a significant genetic component, but clearly smoking, diet, exercise, and many other behavioral and environmental factors all matter, too.

One of the most important contributions of statistical analysis so far has been to debunk false causes, many of which have arisen because of a confusion between correlation and causation. An autism spectrum disorder often appears suddenly between a child's first and second birthdays. This has led to a widespread belief that childhood vaccinations, particularly the triple vaccine for measles, mumps, and rubella (MMR), are causing the rising incidence of autism. Dan Burton, a member of Congress from Indiana, told the *New York Times*, "My grandson received nine shots in one day, seven of which contained thimerosal, which is 50 percent mercury as you know, and he became autistic a short time later."[11]

Scientists have soundly refuted the false association between thimerosal and ASDs. Autism rates did not decline when thimerosal was removed from the MMR vaccine, nor are autism rates lower in countries that never used this vaccine. Nonetheless, the false connection persists, which has caused some parents to refuse to vaccinate their children. Ironically, this offers no protection against autism while putting children at risk of contracting other serious diseases (and contributing to the spread of those diseases in the population).

Autism poses one of the greatest medical and social challenge of our day. We understand so little about the disorder relative to its huge (and possibly growing) impact on our collective well-being. Researchers are using every tool in this book (and lots more) to change that.

HOW CAN WE IDENTIFY AND REWARD GOOD TEACHERS AND SCHOOLS?

We need good schools. And we need good teachers in order to have good schools. Thus, it follows logically that we ought to reward good teachers and good schools while firing bad teachers and closing bad schools.

How exactly do we do that?

Test scores give us an objective measure of student performance. Yet we know that some students will do much better on standardized tests than others for reasons that have nothing to do with what is going on inside a classroom or a school. The seemingly simple solution is to evaluate schools and teachers on the basis of the *progress* that their students make over some period of time. What did students know when they started in a certain classroom with a particular teacher? What did they know a year later? The difference is the "value added" in that classroom.

We can even use statistics to get a more refined sense of this value added by taking into account the demographic characteristics of the students in a given classroom, such as race, income, and performance on other tests (which can be a measure of aptitude). If a teacher makes significant gains with students who have typically struggled in the past, then he or she can be deemed as highly effective.

Voilà! We can now evaluate teacher quality with statistical precision. And the good schools, of course, are just the ones full of effective teachers.

How do these handy statistical evaluations work in practice? In 2012, New York City took the plunge and published ratings of all 18,000 public school teachers on the basis of a "value-added assessment" that measured gains in their students' test scores while taking into account various student characteristics.[12] The *Los Angeles Times* published a similar set of rankings for Los Angeles teachers in 2010.

In both New York and LA, the reaction has been loud and mixed. Arne Duncan, the U.S. secretary of education, has generally been supportive of these kinds of value-added assessments. They provide information where none previously existed. After the Los Angeles data were published, Secretary Duncan told the *New York Times*, "Silence is not an option." The Obama administration has provided financial incentives for states to develop value-added indicators for paying and promoting teachers. Proponents of these evaluation measures rightfully point out that they are a huge potential improvement over systems in which all teachers are paid according to a uniform salary schedule that gives zero weight to any measure of performance in the classroom.

On the other hand, many experts have warned that these kinds of teacher assessment data have large margins of error and can deliver misleading results. The union representing New York City teachers spent more than $100,000 on a newspaper advertising campaign built around the headline "This Is No Way to Rate a Teacher."[13] Opponents argue that the value-added assessments create false precision that will be abused by parents and public officials who do not understand the limitations of this kind of assessment.

This appears to be a case where everybody is right—up to a point. Doug Staiger, an economist at Dartmouth College who works extensively with value-added data for teachers, warns that these data are inherently "noisy." The results for a given teacher are often based on a single test taken on a single day by a single group of students. All kinds of factors can lead to random fluctuations—anything from a particularly difficult group of students to a broken air-conditioning unit clanking away in the classroom on test day. The correlation in performance from year to year for a single teacher that uses these indicators is only about .35. (Interestingly, the correlation in year-to-year performance for Major League baseball players is also around .35, as measured by batting average for hitters and earned run average for pitchers.)[14]

The teacher effectiveness data are useful, says Staiger, but they are just one tool in the process for evaluating teacher performance. The data get "less noisy" when authorities have more years of data for a particular teacher with different classrooms of students (just as we can tell more about an athlete when we have data for more games and more seasons). In the case of the New York City teacher ratings, principals in the system had been prepped on the appropriate use of the value-added data and the inherent limitations. The public did not get that briefing. As a result, the teacher assessments are too often viewed as a definitive guide to the "good" and "bad" teachers. We like rankings—just think *U.S. News & World Report* college rankings—even when the data do not support such precision.

Staiger offers a final warning of different sort: We had better be certain that the outcomes we are measuring, such as the results of a given standardized test, truly track with what we care about in the long run. Some unique data from the Air Force Academy suggest, not surprisingly, that the test scores that glimmer now may not be gold in the future. The Air Force Academy, like the other military academies, randomly assigns its cadets to different sections of standardized core courses, such as introductory calculus. This randomization eliminates any potential selection effect when comparing the effectiveness of professors; over time, we can assume that all professors get students with similar aptitudes (unlike most universities, where students of different abilities can select into or out of different courses). The Air Force Academy also uses the same syllabi and exams in every section of a particular course. Scott Carrell and James West, professors at the University of California at Davis and the Air Force Academy, exploited this elegant arrangement to answer one of the most important questions in higher education: Which professors are most effective?[15]

The answer: *The professors with less experience and fewer degrees from fancy universities*. These professors have students who typically do better on the standardized exams for the introductory courses. They also get better student evaluations for their courses. Clearly these young, motivated instructors are more committed to their teaching than the old, crusty professors with PhDs from places like Harvard. The old guys must be using the same yellowing teaching notes that they used in 1978; they probably think PowerPoint is an energy drink—except that they don't know what an energy drink is either. Obviously the data tell us that we should fire these old codgers, or at least let them retire gracefully.

But hold on. Let's not fire anybody yet. The Air Force Academy study had another relevant finding—about student performance *over a longer horizon*. Carrell and West found that in math and science the students who had more experienced (and more highly credentialed) instructors in the introductory courses *do better in their mandatory follow-on courses* than students who had less experienced professors in the introductory courses. One logical interpretation is that less experienced instructors are more likely to "teach to the test" in the introductory course. This produces impressive exam scores and happy students when it comes to filling out the instructor evaluation.

Meanwhile, the old, crusty professors (whom we nearly fired just one paragraph ago) focus less on the exam and more on the important concepts, which are what matter most in follow-on courses and in life after the Air Force Academy.

Clearly we need to evaluate teachers and professors. We just have to make sure that we do it right. The long-term policy challenge, rooted in statistics, is to develop a system that rewards a teacher's real value added in the classroom.

WHAT ARE THE BEST TOOLS FOR FIGHTING GLOBAL POVERTY?

We know strikingly little about how to make poor countries less poor. True, we understand the things that distinguish rich countries from poor countries, such as their education levels and the quality of their governments. And it is also true that we have watched countries like India and China transform themselves economically over the last several decades.

But even with this knowledge, it is not obvious what steps we can take to make places like Mali or Burkina Faso, less poor. Where should we begin?

French economist Esther Duflo is transforming our knowledge of global poverty by retrofitting an old tool for new purposes: the randomized, controlled experiment. Duflo, who teaches at MIT, literally conducts experiments on different interventions to improve the lives of the poor in developing countries. For example, one of the longstanding problems with schools in India is absenteeism among teachers, particularly in small, rural schools with only a single teacher. Duflo and her coauthor Rema Hanna tested a clever, technology-driven solution on a random sample of 60 one-teacher schools in the Indian state of Rajasthan.[16] Teachers in these 60 experimental schools were offered a bonus for good attendance. Here is the creative part: The teachers were given cameras with tamper-proof date and time stamps. They proved that they had showed up each day by having their picture taken with their students.[17]

Absenteeism dropped by half among teachers in the experimental schools compared with teachers in a randomly selected control group of 60 schools. Student test scores went up, and more students graduated into the next level of education. (I bet the photos are adorable, too!)

One of Duflo's experiments in Kenya involved giving a randomly selected group of farmers a small subsidy to buy fertilizer right *after* the harvest. Prior evidence suggested that fertilizer raises crop yields appreciably. Farmers were aware of this benefit, but when it came time to put a new crop into the ground, they often did not have enough money left over from the last crop to buy fertilizer. This perpetuates what is known as a "poverty trap" since the subsistence farmers are too poor to make themselves less poor. Duflo and her coauthors found that a tiny subsidy—free fertilizer delivery—offered to farmers when they still had cash after the harvest increased fertilizer use by 10 to 20 percentage points compared with use in a control group.[18]

Esther Duflo has even waded into the gender war. Who is more responsible when it comes to handling the family's finances, men or women? In rich countries, this is the kind of thing that couples can squabble over in marriage counseling. In poor countries, it can literally determine whether the children get enough to eat. Anecdotal evidence going back to the dawn of civilization suggests that women place a high priority on the health and welfare of their children, while men are more inclined to drink up their wages at the local pub (or whatever the caveman equivalent was). At worst, this anecdotal evidence merely reinforces age-old stereotypes. At best, it is a hard thing to prove, because a family's finances are comingled to some extent. How can we separate out how husbands and wives choose to spend communal resources?

Duflo did not shy away from this delicate question.[19] To the contrary, she found a fascinating natural experiment. In Côte d'Ivoire, women and men in a family typically share responsibility for some crops. For longstanding cultural reasons, men and women also cultivate different cash crops of their own. (Men grow cocoa, coffee, and some other things; women grow plantains, coconuts, and a few other crops.) The beauty of this arrangement from a research standpoint is that the men's crops and the women's crops respond to rainfall patterns in different ways. In years in which cocoa and coffee do well, men have more disposable income to spend. In years in which plantains and coconuts do well, the women have more extra cash.

Now we need merely broach a delicate question: Are the children in these families better-off in years in which the men's crops do well, or in the years when the women have a particularly bountiful harvest?

The answer: When the women do well, they spend some of their extra cash on more food for the family. The men don't. Sorry guys.

In 2010, Duflo was awarded the John Bates Clark Medal. This prize is presented by the American Economic Association to the best economist under the age of forty.* Among economist geeks, this prize is considered to be more prestigious than the Nobel Prize in Economics because it was historically awarded only every two years. (Beginning with Duflo's award in 2010, the medal is now presented annually.) In any case, the Clark Medal is the MVP award for people with thick glasses (metaphorically speaking).

Duflo is doing program evaluation. Her work, and the work of

---

* I was ineligible for the 2010 prize since I was over forty. Also, I'd done nothing to deserve it.

others now using her methods, is literally changing the lives of the poor. From a statistical standpoint, Duflo's work has encouraged us to think more broadly about how randomized, controlled experiments—long thought to be the province of the laboratory sciences—can be used more widely to tease out causal relationships in many other areas of life.

WHO GETS TO KNOW WHAT ABOUT YOU?

Last summer, we hired a new babysitter. When she arrived at the house, I began to explain our family background: "I am a professor, my wife is a teacher . . ."

"Oh, I know," the sitter said with a wave of the hand. "I Googled you."

I was simultaneously relieved that I did not have to finish my spiel and mildly alarmed by how much of my life could be cobbled together from a short Internet search. Our capacity to gather and analyze huge quantities of data—the marriage of digital information with cheap computing power and the Internet—is unique in human history. We are going to need some new rules for this new era.

Let's put the power of data in perspective with just one example from the retailer Target. Like most companies, Target strives to increase profits by understanding its customers. To do that, the company hires statisticians to do the kind of "predictive analytics" described earlier in the book; they use sales data combined with other information on consumers to figure out who buys what and why. Nothing about this is inherently bad, for it means that the Target near you is likely to have exactly what you want.

But let's drill down for a moment on just one example of the kinds of things that the statisticians working in the windowless basement at corporate headquarters can figure out. Target has learned that pregnancy is a particularly important time in terms of developing shopping patterns. Pregnant women develop "retail relationships" that can last for decades. As a result, Target wants to identify pregnant women, particularly those in their second trimester, and get them into their stores more often. A writer for the *New York Times Magazine* followed the predictive analytics team at Target as it sought to find and attract pregnant shoppers.[20]

The first part is easy. Target has a baby shower registry in which pregnant women register for baby gifts in advance of the birth of their children. These women are already Target shoppers, and they've effectively told the store that they are pregnant. But here is the statistical twist: *Target figured out that other women who demonstrate the same shopping patterns are probably pregnant, too.* For example, pregnant women often switch to unscented lotions. They begin to buy vitamin supplements. They start buying extrabig bags of cotton balls. The Target predictive analytics gurus identified twenty-five products that together made possible a "pregnancy prediction score." The whole point of this analysis was to send pregnant women pregnancy-related coupons in hopes of hooking them as long-term Target shoppers.

How good was the model? The *New York Times Magazine* reported a story about a man from Minneapolis who walked into a Target store and demanded to see a manager. The man was irate that his high school daughter was being bombarded with pregnancy-related coupons from Target. "She's still in high school and you're sending her coupons for baby clothes and cribs? Are you trying to encourage her to get pregnant?" the man asked.

The store manager apologized profusely. He even called the father several days later to apologize again. Only this time, the man was less irate; it was his turn to be apologetic. "It turns out there's been some activities in my house I haven't been completely aware of," the father said. "She's due in August."

The Target statisticians had figured out that his daughter was pregnant before he did.

That is their business . . . and also not their business. It can feel more than a little intrusive. For that reason, some companies now mask how much they know about you. For example, if you are a pregnant woman in your second trimester, you may get some coupons in the mail for cribs and diapers—along with a discount on a riding lawn mower and a coupon for free bowling socks with the purchase of any pair of bowling shoes. To you, it just seems fortuitous that the pregnancy-related coupons came in the mail along with the other junk. In fact, the company knows that you don't bowl or cut your own lawn; it's merely covering its tracks so that what it knows about you doesn't seem so spooky.

Facebook, a company with virtually no physical assets, has become one of the most valuable companies in the world. To investors (as opposed to users), Facebook has one enormous asset: data. Investors don't love Facebook because it allows them to reconnect with their prom dates. They love Facebook because every click of the mouse yields data about where users live, where they shop, what they buy, who they know, and how they spend their time. To users, who *are* hoping to reconnect with their prom dates, the corporate data gathering can overstep the boundaries of privacy.

Chris Cox, Facebook's vice president of product, told the *New York Times*, "The challenge of the information age is what to do with it."[21]

Yep.

And in the public arena, the marriage of data and technology gets even trickier. Cities around the world have installed thousands of security cameras in public places, some of which will soon have facial recognition technology. Law enforcement authorities can follow any car anywhere it may go (and keep extensive records of where it has been) by attaching a global positioning device to the vehicle and then tracking it by satellite. Is this a cheap and efficient way to monitor potential criminal activity? Or is this the government using technology to trample on our personal liberty? In 2012, the U.S. Supreme Court decided unanimously that it was the latter, ruling that law enforcement officials can no longer attach tracking devices to private vehicles without a warrant.*

Meanwhile, governments around the world maintain huge DNA databases that are a powerful tool for solving crimes. Whose DNA should be in the database? That of all convicted criminals? That of every person who is arrested (whether or not eventually convicted)? Or a sample from every one of us?

We are just beginning to wrestle with the issues that lie at the inter-

---

* *The United States v. Jones.*

section of technology and personal data—none of which were terribly relevant when government information was stored in dusty basement filing cabinets rather than in digital databases that are potentially searchable by anyone from anywhere. Statistics is more important than ever before because we have more meaningful opportunities to make use of data. Yet the formulas will not tell us which uses of data are appropriate and which are not. Math cannot supplant judgment.

In that vein, let's finish the book with some word association: fire, knives, automobiles, hair removal cream. Each one of these things serves an important purpose. Each one makes our lives better. And each one can cause some serious problems when abused.

Now you can add statistics to that list. Go forth and use data wisely and well!

Appendix
Statistical software

I suspect that you won't be doing your statistical analysis with a pencil, paper, and calculator. Here is a quick tour of the software packages most commonly used for the kinds of tasks described in this book.

Microsoft Excel

Microsoft Excel is probably the most widely used program to compute simple statistics such as mean and standard deviation. Excel can also do basic regression analysis. Most computers come loaded with Microsoft Office, so Excel is probably sitting on your desk right now. Excel is user-friendly compared with more sophisticated statistical software packages. The basic statistical calculations can be done by means of the formula bar.

Excel cannot perform some of the advanced tasks that more specialized programs can do. However, there are Excel extensions that you can buy (and some that you can download for free) that will expand the program's statistical capabilities. One huge advantage to Excel is that it offers simple ways to display two-dimensional data with visually appealing graphics. These graphics can be easily dropped into Microsoft PowerPoint and Microsoft Word.

*Stata**

Stata is a statistical package used worldwide by research professionals; its interface has a serious, academic feel. Stata has a wide range of capabilities to do basic tasks, such as creating data tables and calculating descriptive statistics. Of course, that is not why university professors and other serious researchers choose Stata. The software is designed to handle sophisticated statistical tests and data modeling that are far beyond the kinds of things described in this book.

Stata is a great fit for those who have a solid understanding of statistics (a basic understanding of programming also helps) and those who do not need fancy formatting—just the answers to their statistical queries. Stata is not the best choice if your goal is to produce quick graphics from the data. Expert users say that Stata can produce nice graphics but that Excel is easier to use for that purpose.

Stata offers several different stand-alone software packages. You can either license the product for a year (after a year, the software no longer works on your computer) or license it forever. One of the cheapest options is Stata/IC, which is designed for "students and researchers with moderate-sized datasets." There is a discount for users who are in the education sector. Even then, a single-user annual license for Stata/IC is $295 and a perpetual license is $595. If you plan to launch a satellite to Mars and need to do some really serious number crunching, you can look into more advanced Stata packages, which can cost thousands of dollars.

SAS†

SAS has a broad appeal not only to professional researchers but also to business analysts and engineers because of its broad range of analyti-

* See http://www.stata.com/.

† See http://www.sas.com/technologies/analytics/statistics/.

cal capabilities. SAS sells two different statistical packages. The first is called SAS Analytics Pro, which can read data in virtually any format and perform advanced data analysis. The software also has good data visualization tools, such as advanced mapping capabilities. It's not cheap. Even for those in the education and government sectors, a single commercial or individual license for this package is $8,500, plus an annual license fee.

The second SAS statistical package is SAS Visual Data Discovery. It has an easy-to-use interface that requires no knowledge of coding or programming, while still providing advanced data analysis capabilities. As its name suggests, this package is meant to allow the user to easily explore data with interactive visualization. You can also export the data animations into presentations, Web pages, and other documents. This one is not cheap either. A single commercial or individual license for this package is $9,810, plus an annual license fee.

SAS sells some specialized management tools, such as a product that uses statistics to detect fraud and financial crimes.

R

This may sound like a character in a James Bond movie. In fact, R is a popular statistical package that is free or "open source." It can be downloaded and easily installed on your computer in a matter of minutes. There is also an active "R community" that shares code and can offer help and guidance when needed.

Not only is R the cheapest option, but it is also one of the most malleable of all of the packages described here. Depending on your perspective, this flexibility is either frustrating or one of R's great assets. If you are new to statistical software, the program offers almost no structure. The interface will not help you along much. On the other hand, programmers (and even people who have just a basic familiarity with coding principles) can find the lack of structure liberating. Users are free to tell the program to do exactly what they want it to do, including having it work with outside programs.

*IBM SPSS**

IBM SPSS has something for everyone, from hard-core statisticians to less statistically rugged business analysts. IBM SPSS is good for beginners because it offers a menu-driven interface. IBM SPSS also offers a range of tools or "modules" that are designed to perform specific functions, such as IBM SPSS Forecasting, IBM SPSS Advanced Statistics, IBM SPSS Visualization Designer, and IBM SPSS Regression. The modules can be purchased individually or combined into packages.

The most basic package offered is IBM SPSS Statistics Standard Edition, which allows you to calculate simple statistics and perform basic data analysis, such as identifying trends and building predictive models. A single fixed-term commercial license is $2,250. The premium package, which includes most of the modules, is $6,750. Discounts are available for those who work in the education sector.

* See http://www-01.ibm.com/software/analytics/spss/products/statistics/.

Notes

Chapter 1: What's the Point?

1 Central Intelligence Agency, *The World Factbook*, https://www.cia.gov/library/publications/the-world-factbook/.
2 Steve Lohr, "For Today's Graduate, Just One Word: Statistics," *New York Times*, August 6, 2009.
3 Ibid.
4 Baseball-Reference.com, http://www.baseball-reference.com/players/m/mantlmi01.shtml.
5 Trip Gabriel, "Cheats Find an Adversary in Technology," *New York Times*, December 28, 2010.
6 Eyder Peralta, "Atlanta Man Wins Lottery for Second Time in Three Years," *NPR News* (blog), November 29, 2011.
7 Alan B. Krueger, *What Makes a Terrorist: Economics and the Roots of Terrorism* (Princeton: Princeton University Press, 2008).

Chapter 2: Descriptive Statistics

1 U.S. Census Bureau, Current Population Survey, Annual Social and Economic Supplements, http://www.census.gov/hhes/www/income/data/historical/people/.
2 Malcolm Gladwell, "The Order of Things," *The New Yorker*, February 14, 2011.
3 CIA, *World Factbook*, and United Nations Development Program, *2011 Human Development Report*, http://hdr.undp.org/en/statistics/.
4 Baseball-Reference.com.

Chapter 3: Deceptive Description

1 Robert Griffith, *The Politics of Fear: Joseph R. McCarthy and the Senate*, 2nd ed. (Amherst: University of Massachusetts Press, 1987), p. 49.
2 "Catching Up," *Economist*, August 23, 2003.
3 Carl Bialik, "When the Median Doesn't Mean What It Seems," *Wall Street Journal*, May 21–22, 2011.

4 Stephen Jay Gould, "The Median Isn't the Message," with a prefatory note and postscript by Steve Dunn, http://cancerguide.org/median_not_msg.html.
5 See http://www.movieweb.com/box-office/alltime.
6 Box Office Mojo (boxofficemojo.com), June 29, 2011.
7 Steve Patterson, "527% Tax Hike May Shock Some, But It's Only About $5," *Chicago Sun-Times*, December 5, 2005.
8 Rebecca Leung, "'The 'Texas Miracle': *60 Minutes II* Investigates Claims That Houston Schools Falsified Dropout Rates," CBSNews.com, August 25, 2004.
9 Marc Santora, "Cardiologists Say Rankings Sway Surgical Decisions," *New York Times*, January 11, 2005.
10 Interview with National Public Radio, August 20, 2006, http://www.npr.org/templates/story/story.php?storyId=5678463.
11 See http://www.usnews.com/education/articles/2010/08/17/frequently-asked-questions-college-rankings#4.
12 Gladwell, "Order of Things."
13 Interview with National Public Radio, February 22, 2007, http://www.npr.org/templates/story/story.php?storyId=7383744.

Chapter 4: Correlation

1 College Board, FAQs, http://www.collegeboard.com/prod_downloads/about/news_info/cbsenior/yr2010/correlations-of-predictors-with-first-year-college-grade-point-average.pdf.
2 College Board, 2011 College-Bound Seniors Total Group Profile Report, http://professionals.collegeboard.com/profdownload/cbs2011_total_group_report.pdf.
3 See http://www.netflixprize.com/rules.

Chapter 5: Basic Probability

1 David A. Aaker, *Managing Brand Equity: Capitalizing on the Value of a Brand Name* (New York: Free Press, 1991).
2 Victor J. Tremblay and Carol Horton Tremblay, *The U.S. Brewing Industry: Data and Economic Analysis* (Cambridge: MIT Press, 2005).
3 Australian Transport Safety Bureau Discussion Paper, "Cross Modal Safety Comparisons," January 1, 2005.
4 Marcia Dunn, "1 in 21 Trillion Chance Satellite Will Hit You," *Chicago Sun-Times*, September 21, 2011.
5 Steven D. Levitt and Stephen J. Dubner, *Freakonomics: A Rogue Economist Explores the Hidden Side of Everything* (New York: William Morrow Paperbacks, 2009).

6 Garrick Blalock, Vrinda Kadiyali, and Daniel Simon, "Driving Fatalities after 9/11: A Hidden Cost of Terrorism" (unpublished manuscript, December 5, 2005).
7 The general genetic testing information comes from Human Genome Project Information, DNA Forensics, http://www.ornl.gov/sci/techresources/Human_Genome/elsi/forensics.shtml.
8 Jason Felch and Maura Dolan, "FBI Resists Scrutiny of 'Matches,' " *Los Angeles Times*, July 20, 2008.
9 David Leonhardt, "In Football, 6 + 2 Often Equals 6," *New York Times*, January 16, 2000.
10 Roger Lowenstein, "The War on Insider Trading: Market Beaters Beware," *New York Times Magazine*, September 22, 2011.
11 Erica Goode, "Sending the Police before There's a Crime," *New York Times*, August 15, 2011.
12 The insurance risk data come from all of the following: "Teen Drivers," Insurance Information Institute, March 2012; "Texting Laws and Collision Claim Frequencies," Insurance Institute for Highway Safety, September 2010; "Hot Wheels," National Insurance Crime Bureau, August 2, 2011.
13 Charles Duhigg, "What Does Your Credit Card Company Know about You?" *New York Times Magazine*, May 12, 2009.

Chapter 5½: The Monty Hall Problem

1 John Tierney, "And behind Door No. 1, a Fatal Flaw," *New York Times*, April 8, 2008.
2 Leonard Mlodinow, *The Drunkard's Walk: How Randomness Rules Our Lives* (New York: Vintage Books, 2009).

Chapter 6: Problems with Probability

1 Joe Nocera, "Risk Mismanagement," *New York Times Magazine*, January 2, 2009.
2 Robert E. Hall, "The Long Slump," *American Economic Review* 101, no. 2 (April 2011): 431–69.
3 Alan Greenspan, Testimony before the House Committee on Government Oversight and Reform, October 23, 2008.
4 Hank Paulson, Speech at Dartmouth College, Hanover, NH, August 11, 2011.
5 "The Probability of Injustice," *Economist*, January 22, 2004.
6 Thomas Gilovich, Robert Vallone, and Amos Tversky, "The Hot Hand in Basketball: On the Misperception of Random Sequences," *Cognitive Psychology* 17, no. 3 (1985): 295–314.

7 Ulrike Malmendier and Geoffrey Tate, "Superstar CEOs," *Quarterly Journal of Economics* 124, no. 4 (November 2009): 1593–638.
8 "The Price of Equality," *Economist*, November 15, 2003.

Chapter 7: The Importance of Data

1 Benedict Carey, "Learning from the Spurned and Tipsy Fruit Fly," *New York Times*, March 15, 2012.
2 Cynthia Crossen, "Fiasco in 1936 Survey Brought 'Science' to Election Polling," *Wall Street Journal*, October 2, 2006.
3 Tara Parker-Pope, "Chances of Sexual Recovery Vary Widely after Prostate Cancer," *New York Times*, September 21, 2011.
4 Benedict Carey, "Researchers Find Bias in Drug Trial Reporting," *New York Times*, January 17, 2008.
5 Siddhartha Mukherjee, "Do Cellphones Cause Brain Cancer?" *New York Times*, April 17, 2011.
6 Gary Taubes, "Do We Really Know What Makes Us Healthy?" *New York Times*, September 16, 2007.

Chapter 8: The Central Limit Theorem

1 U.S. Census Bureau.

Chapter 9: Inference

1 John Friedman, *Out of the Blue: A History of Lightning: Science, Superstition, and Amazing Stories of Survival* (New York: Delacorte Press, 2008).
2 "Low Marks All Round," *Economist*, July 14, 2011.
3 Trip Gabriel and Matt Richtel, "Inflating the Software Report Card," *New York Times*, October 9, 2011.
4 Jennifer Corbett Dooren, "Link in Autism, Brain Size," *Wall Street Journal*, May 3, 2011.
5 Heather Cody Hazlett et al., "Early Brain Overgrowth in Autism Associated with an Increase in Cortical Surface Area before Age 2 Years," *Archives of General Psychiatry* 68, no. 5 (May 2011): 467–76.
6 Benedict Carey, "Top Journal Plans to Publish a Paper on ESP, and Psychologists Sense Outrage," *New York Times*, January 6, 2011.

Chapter 10: Polling

1 Jeff Zeleny and Megan Thee-Brenan, "New Poll Finds a Deep Distrust of Government," *New York Times*, October 26, 2011.

2 Lydia Saad, "Americans Hold Firm to Support for Death Penalty," Gallup.com, November 17, 2008.
3 Phone interview with Frank Newport, November 30, 2011.
4 Stanley Presser, "Sex, Samples, and Response Errors," *Contemporary Sociology* 24, no. 4 (July 1995): 296–98.
5 The results were published in two different formats, one more academic than the other. Edward O. Lauman, *The Social Organization of Sexuality: Sexual Practices in the United States* (Chicago: University of Chicago Press, 1994); Robert T. Michael, John H. Gagnon, Edward O. Laumann, and Gina Kolata, *Sex in America: A Definitive Survey* (New York: Grand Central Publishing, 1995).
6 Kaye Wellings, book review in *British Medical Journal* 310, no. 6978 (February 25, 1995): 540.
7 John DeLamater, "The NORC Sex Survey," *Science* 270, no. 5235 (October 20, 1995): 501.
8 Presser, "Sex, Samples, and Response Errors."

Chapter 11: Regression Analysis

1 Marianne Bertrand, Claudia Goldin, and Lawrence F. Katz, "Dynamics of the Gender Gap for Young Professionals in the Corporate and Financial Sectors," NBER Working Paper 14681, January 2009.
2 M. G. Marmot, Geoffrey Rose, M. Shipley, and P. J. S. Hamilton, "Employment Grade and Coronary Heart Disease in British Civil Servants," *Journal of Epidemiology and Community Health* 32, no. 4 (1978): 244–49.
3 Hans Bosma, Michael G. Marmot, Harry Hemingway, Amanda C. Nicholson, Eric Brunner, and Stephen A. Stansfeld, "Low Job Control and Risk of Coronary Heart Disease in Whitehall II (Prospective Cohort) Study," *British Medical Journal* 314, no. 7080 (February 22, 1997): 558–65.
4 Peter L. Schnall, Paul A. Landesbergis, and Dean Baker, "Job Strain and Cardiovascular Disease," *Annual Review of Public Health* 15 (1994): 381–411.
5 M. G. Marmot, H. Bosma, H. Hemingway, E. Brunner, and S. Stansfeld, "Contribution of Job Control and Other Risk Factors to Social Variations in Coronary Heart Disease Incidence," *Lancet* 350 (July 26, 1997): 235–39.

Chapter 12: Common Regression Mistakes

1 Gary Taubes, "Do We Really Know What Makes Us Healthy?" *New York Times Magazine*, September 16, 2007.
2 "Vive la Difference," *Economist*, October 20, 2001.
3 Taubes, "Do We Really Know?"
4 College Board, 2011 College-Bound Seniors Total Group Profile Report, http://professionals.collegeboard.com/profdownload/cbs2011_total_group_report.pdf.

5 Hans Bosma et al., "Low Job Control and Risk of Coronary Heart Disease in Whitehall II (Prospective Cohort) Study," *British Medical Journal* 314, no. 7080 (February 22, 1997): 564.
6 Taubes, "Do We Really Know?"
7 Gautam Naik, "Scientists' Elusive Goal: Reproducing Study Results," *Wall Street Journal*, December 2, 2011.
8 John P. A. Ioannidis, "Contradicted and Initially Stronger Effects in Highly Cited Clinical Research," *Journal of the American Medical Association* 294, no. 2 (July 13, 2005): 218–28.
9 "Scientific Accuracy and Statistics," *Economist*, September 1, 2005.

Chapter 13: Program Evaluation

1 Gina Kolata, "Arthritis Surgery in Ailing Knees Is Cited as Sham," *New York Times*, July 11, 2002.
2 Benedict Carey, "Long-Awaited Medical Study Questions the Power of Prayer," *New York Times*, March 31, 2006.
3 Diane Whitmore Schanzenbach, "What Have Researchers Learned from Project STAR?" Harris School Working Paper, August 2006.
4 Gina Kolata, "A Surprising Secret to a Long Life: Stay in School," *New York Times*, January 3, 2007.
5 Adriana Lleras-Muney, "The Relationship between Education and Adult Mortality in the United States," *Review of Economic Studies* 72, no. 1 (2005): 189–221.
6 Kurt Badenhausen, "Top Colleges for Getting Rich," Forbes.com, July 30, 2008.
7 Stacy Berg Dale and Alan Krueger, "Estimating the Payoff to Attending a More Selective College: An Application of Selection on Observables and Unobservables," *Quarterly Journal of Economics* 117, no. 4 (November 2002): 1491–527.
8 Alan B. Krueger, "Children Smart Enough to Get into Elite Schools May Not Need to Bother," *New York Times*, April 27, 2000.
9 Randi Hjalmarsson, "Juvenile Jails: A Path to the Straight and Narrow or to Hardened Criminality?" *Journal of Law and Economics* 52, no. 4 (November 2009): 779–809.

Conclusion

1 James Surowiecki, "A Billion Prices Now," *The New Yorker*, May 30, 2011.
2 Malcolm Gladwell, "Offensive Play," *The New Yorker*, October 19, 2009.
3 Ken Belson, "N.F.L. Roundup; Concussion Suits Joined," *New York Times*, February 1, 2012.
4 Shirley S. Wang, "Autism Diagnoses Up Sharply in U.S.," *Wall Street Journal*, March 30, 2012.

5 Catherine Rice, "Prevalence of Autism Spectrum Disorders," Autism and Developmental Disabilities Monitoring Network, Centers for Disease Control and Prevention, 2006, http://www.cdc.gov/mmwr/preview/mmwrhtml/ss5810a1.htm.

6 Alan Zarembo, "Autism Boom: An Epidemic of Disease or of Discovery?" latimes.com, December 11, 2011.

7 Michael Ganz, "The Lifetime Distribution of the Incremental Societal Costs of Autism," *Archives of Pediatrics & Adolescent Medicine* 161, no. 4 (April 2007): 343–49.

8 Gardiner Harris and Anahad O'Connor, "On Autism's Cause, It's Parents vs. Research," *New York Times*, June 25, 2005.

9 Julie Steenhuysen, "Study Turns Up 10 Autism Clusters in California," *Yahoo! News*, January 5, 2012.

10 Joachim Hallmayer et al., "Genetic Heritability and Shared Environmental Factors among Twin Pairs with Autism," *Archives of General Psychiatry* 68, no. 11 (November 2011): 1095–102.

11 Gardiner Harris and Anahad O'Connor, "On Autism's Cause, It's Parents vs. Research," *New York Times*, June 25, 2005.

12 Fernanda Santos and Robert Gebeloff, "Teacher Quality Widely Diffused, Ratings Indicate," *New York Times*, February 24, 2012.

13 Winnie Hu, "With Teacher Ratings Set to Be Released, Union Opens Campaign to Discredit Them," *New York Times*, February 23, 2012.

14 T. Schall and G. Smith, "Do Baseball Players Regress to the Mean?" *American Statistician* 54 (2000): 231–35.

15 Scott E. Carrell and James E. West, "Does Professor Quality Matter? Evidence from Random Assignment of Students to Professors," National Bureau of Economic Research Working Paper 14081, June 2008.

16 Esther Duflo and Rema Hanna, "Monitoring Works: Getting Teachers to Come to School," National Bureau of Economic Research Working Paper 11880, December 2005.

17 Christopher Udry, "Esther Duflo: 2010 John Bates Clark Medalist," *Journal of Economic Perspectives* 25, no. 3 (Summer 2011): 197–216.

18 Esther Duflo, Michael Kremer, and Jonathan Robinson, "Nudging Farmers to Use Fertilizer: Theory and Experimental Evidence from Kenya," National Bureau of Economic Research Working Paper 15131, July 2009.

19 Esther Duflo and Christopher Udry, "Intrahousehold Resource Allocation in Côte d'Ivoire: Social Norms, Separate Accounts and Consumption Choices," Working Paper, December 21, 2004.

20 Charles Duhigg, "How Companies Learn Your Secrets," *New York Times Magazine*, February 16, 2012.

21 Somini Sengupta and Evelyn M. Rusli, "Personal Data's Value? Facebook Set to Find Out," *New York Times*, February 1, 2012.

Index

Page numbers in *italics* refer to figures.